Network Management with Smart Systems

Other McGraw-Hill Books of Interest

BALL • *Cost-Efficient Network Management*, 0-07-003484-2

BERSON • *APPC*, 0-07-005075-9

BERSON • *Client/Server Architecture*, 0-07-005076-7

BLACK • *Frame Relay Specifications and Implementations*, 0-07-005558-0

BLACK • *Network Management Standards*, 0-07-005554-8

BLACK • *TCP/IP and Related Protocols*, 0-07-005553-X

BLACK • *The V Series Recommendations*, 0-07-005552-1

BLACK • *The X Series Recommendations*, 0-07-005546-7

CERUTTI • *Distributed Computing Environments*, 0-07-010516-2

DAYTON • *Multi-Vendor Networks*, 0-07-016196-8

DEWIRE • *Application Development for Distributed Systems*, 0-07-016733-8

DEWIRE • *Client/Server Computing*, 0-07-016732-X

EDMUNDS • *SAA/LU6.2*, 0-07-019022-4

FEIT • *TCP/IP*, 0-07-020346-6

HEBRAWI • *OSI Upper Layer Standards and Practices*, 0-07-033754-3

HELDMAN • *Future Telecommunications*, 0-07-028039-8

HELDMAN • *Global Telecommunications*, 0-07-028030-4

JAIN • *Open Systems Interconnection*, 0-07-032385-2

KESSLER • *ISDN*, 0-07-034247-4

KESSLER • *Metropolitan Area Networks*, 0-07-034243-1

KNIGHTSON • *OSI Protocol Performance Testing: IS 9646 Explained*, 0-07-035134-1

MINOLI • *1ST, 2ND, and Next Generation LANs*, 0-07-042586-8

NAUGLE • *Local Area Networking*, 0-07-046455-3

NAUGLE • *Network Protocol Handbook*, 0-07-046461-8

NEMZOW • *The Ethernet Management Guide*, 0-07-046320-4

NEMZOW • *The Token-Ring Management Guide*, 0-07-046321-2

RADICATI • *Electronic Mail*, 0-07-051104-7

SACKETT • *IBM's Token-Ring Networking Handbook*, 0-07-054418-2

SPOHN • *Data Network Design*, 0-07-060360-X

TERPLAN • *Effective Management of Local Area Networks*, 0-07-063636-2

To order, or to receive additional information on these or any other McGraw-Hill titles, please call 1-800-822-8158 in the United States. In other countries, contact your local McGraw-Hill office.

MH93

Network Management with Smart Systems

Larry L. Ball, Ph.D.
Managing Director
Telenetic Controls, Ltd.
Vancouver, B.C., Canada

McGraw-Hill, Inc.
New York   San Francisco   Washington, D.C.   Auckland   Bogotá
Caracas   Lisbon   London   Madrid   Mexico City   Milan
Montreal   New Delhi   San Juan   Singapore
Sydney   Tokyo   Toronto

Library of Congress Cataloging-in-Publication Data

Ball, Larry Lennox.
    Network management with smart systems / Larry L. Ball.
        p.   cm. — (McGraw-Hill series on computer communications)
    Includes index.
    ISBN 0-07-003600-4
    1. Computer networks—Management—Data processing. 2. Neural
networks (Computer science) 3. Expert systems (Computer science)
I. Title. II. Series.
TK5105.5.B3423   1994
004.6—dc20   93-8686
    CIP

The sponsoring editor for this book was Jeanne Glasser, the editing supervisor was Caroline Levine, and the production supervisor was Suzanne W. Babeuf. It was composed by the author.

Printed and bound by R. R. Donnelley & Sons Company.

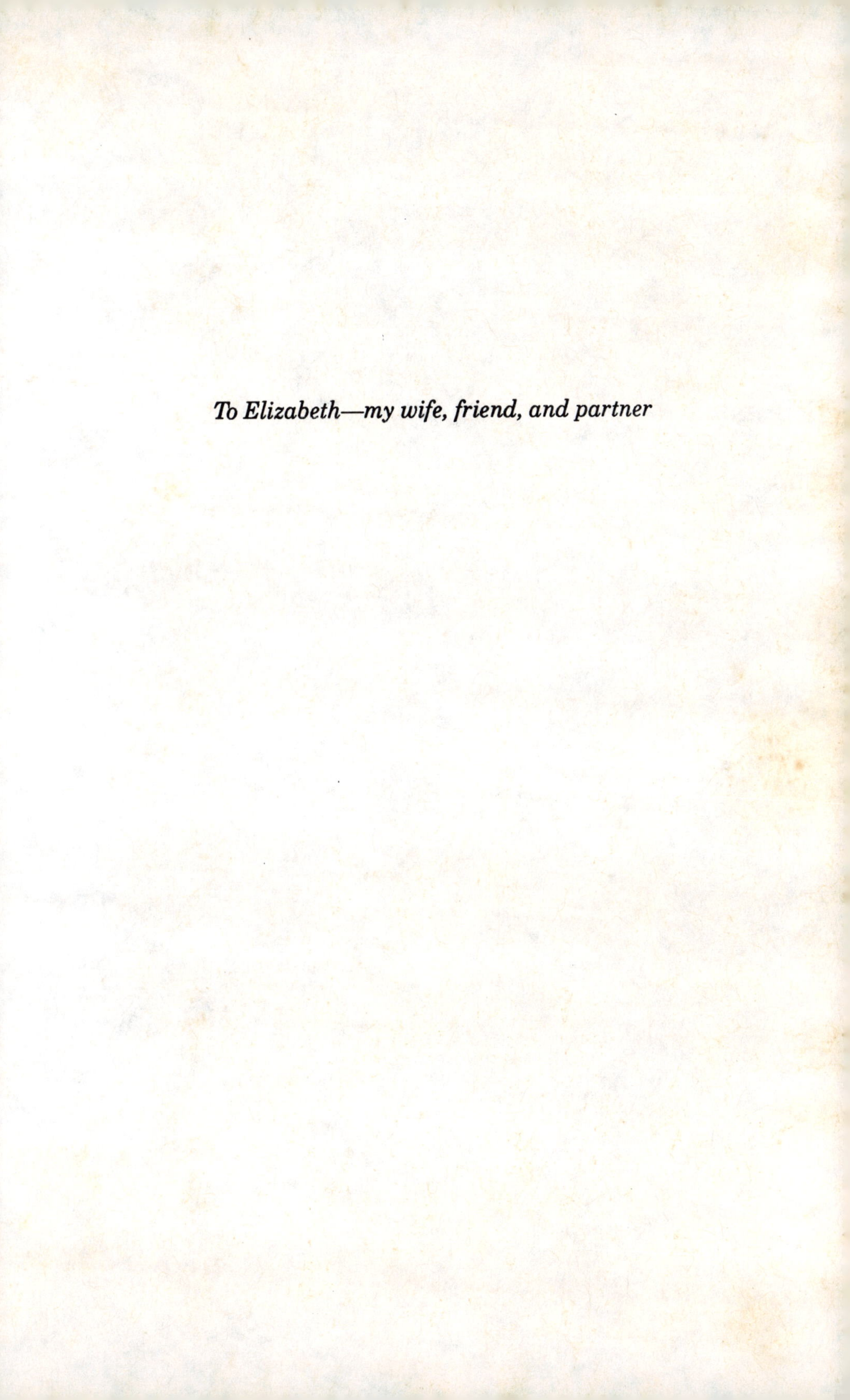

To Elizabeth—my wife, friend, and partner

Contents

Chapter 2. The Functional Requirements of Smart Systems 49

Chapter 3. Sensors and Sensor Systems 79

Chapter 4. Data Fusion Processing Systems 111

Chapter 7. Expert Systems: Maintenance Systems.................................199

Chapter 8. Expert Systems: Performance Systems......239

Chapter 9. Neural Networks: Identifying Their Potential......267

Chapter 10. Neural Networks: Design and Usage

PREFACE

The struggle for world power, influence, and position is shifting from the battlefields to the boardrooms. Countries and multinational corporations are beginning to compete with each other more with economic weapons than with military ones. The technology buildups in military-related areas are now giving way to the transfer of these technologies to the development and marketing of new civilian uses or to more cost-efficient methods of producing familiar products. The current destinations of many of these one-time, military technologies are the corporate product development centers, software laboratories, and hardware manufacturing research facilities. Their strategic values and competitive advantages are being assessed in the marketing and information organizations of large companies, so that financial ventures can be gauged and packaged properly. In short, the battles for survival are shifting from deserts and jungles to homes and shopping malls.

The expertise and systems developed over the past 50 years of global conflict present tremendous opportunities for the uses of these tools to make the now home-front battles swing in the favor of the organization that can understand and react to these advantages. These tools are, for the most part, just lying around, waiting to be utilized. The use of military command and control methodologies, and tools and techniques, in particular, are directly applicable to such endeavors as network management, executive information support, telecommunications control, process control, and the like. This area of human endeavor will probably be the next frontier after the information age has matured.

While it is true that network management has been subjected to rigorous standardization by international groups, it is also true that the competitive advantage for those vendors of such products and services utilized in network management will rest upon the ability of the individual company to differentiate itself by providing features that make its product more desirable than a competitor's product for some reason or other. In many cases, these differentiating features come in the form of unique technologies, or simple, elegant applications of technology solutions. Smart systems provide such an opportunity for differentiation.

Smart systems technologies have been around, in some cases, for many years, however, their implementation as cost-effective solutions has been impaired due to slow technology platforms, costly equipment, embellishment and mismanagement of expectations, and/or erroneous specifications of performance. It took until the mid-1980s for smart systems to become a credible tool in the tool kit of developers and it is just now becoming widely accepted by users.

This book is developed in two parts. The first part reviews some of the basic issues of smart systems involvement in network management, and the command and control aspects of smart systems implementation. The second part develops the specific uses of smart systems and discusses and explores the strengths and weaknesses of different smart systems technologies, such as neural networks, expert systems, etc. Outlined below, are the summaries for chapters in this book.

Part 1

Chapter 1

A review of the relevant issues associated with each of the enabling technologies is presented. These technologies include sensors, data fusion, and decision support. The whole area of smart systems technology is an outgrowth of military systems, in particular the command, control, communication and intelligence (C^3I) technology development work that has occurred during the past 50 years. In addition, however, the evolution of technology itself has greatly influenced the advancement of smart systems work applicable to network management that has currently resulted in capabilities not possible just a few years ago. This chapter surveys the present status of technology related to sensors, data fusion, and decision support.

Chapter 2

The functional requirements related to smart systems usage within the network management environment are explored. Smart systems technology as it relates to network management embraces the precepts and standards associated with the OSI network management model. In this chapter we will review the five major functions associated with the OSI model, and discuss some of the opportunities available to smart systems as they might apply to these five functions. The underlying assumption in this chapter is that smart sensors, decision support systems, and data fusion are all involved in the solutions and opportunities proposed.

Chapter 3 - Sensors and Sensor Systems

Sensor technology has enjoyed an increase in interest and a boost in technological improvement over the past few years. This chapter will trace the status of sensor technology, the types available, the architectures surrounding their configurations, and new developments in this important area. In the world of digital network management, sensors have taken on new forms,as well as new functions that they must perform. The step from analog to digital sensing had left sensor technology playing a catchup game until recent times. Sensors are evolving into communications devices and eventually will be either the main control unit or a backup control unit for the equipment item itself.

Chapter 4 - Data Fusion Processing Systems

Smart systems derive much of their intelligence from the combinational effects of the collection activities supporting the front end of the system operation. Data fusion has a historical basis for its justification, beginning with or possibly even before the writings of Bayes. This chapter is about the development of fusion concepts and their application to today's sophisticated communications environments. There is no magic here. The basic concepts will be used to illustrate the utility of data fusion and its relationships to network management operations.

Chapter 5 - Decision Support Systems

Decision support systems (DSS) are derived from command and control systems developed by the military during the past 50 years. They embody the same features insofar as the data analysis is concerned and, to some extent, include many of the same types of decision making algorithms and data collection processing found in military systems. This chapter addresses these systems in terms of their architectures, assessment capabilities, feature design, and competitive advantage issues.

Part 2

Chapter 6 - Expert Systems: Human Aspects

Artificial intelligence has attempted to emulate certain of the characteristics of the human operator and decision maker. Within this category are such expert systems that are supposed to emulate human thoughts and reasoning. The definition, categorization, and distribution of expertise are discussed in this chapter, along with a generalized description of what an expert system is. The key topics in a chapter that discusses the human aspects of expert systems are those of knowledge engineering and natural language processing, both of which are explored here in detail. The translation of knowledge into something that the computer can deal with and utilize is the ultimate hurdle in any expert system implementation.

Chapter 7 - Expert Systems: Maintenance Systems

Probably the most directly applicable use of expert systems to network management has been in the realm of maintenance operations. Operations and maintenance are the two most critical areas of current-day interests in network management and, as such, their relationship to AI and expert systems influence, in particular, cannot be overemphasized. Expert maintenance systems aid in identifying problems, and isolating their location. Provided in this chapter are the fundamentals of the critical and cost-saving technology of this decade, if not beyond.

Chapter 8 - Expert Systems: Performance Systems

Performance monitoring and control will be a fertile area for expert systems influence in the network management arena. These systems are aimed at improving overall operational efficiencies, particularly in the areas of traffic management and network control. This chapter explores the tools used by the designer to create expert-based performance systems.

Chapter 9 - Neural Networks: Identifying Their Potential

Neural network technology mimics its biological antecedent. Investigations of the vertebrate nervous system have revealed architectures and processing methods that, initially were simulated on computers, and later were borrowed to develop new approaches related to the whole issue of computerized adaptation and learning. Parallel computer systems and distributed computer systems have been the principal mechanisms allowing this technology to flourish. Massively parallel systems incorporate many processing nodes, each of which is its own computer and is used as a correlate for a neuron. This chapter surveys the basic knowledge about the neuron and some of the learning, memory, as well as applications associated with neural networks.

Chapter 10 - Neural Networks: Design and Usage

The design and architectural issues associated with neural networks are the key to their utility in network management environments. Neural networks are presently enjoying some initial success in their implementation in certain limited applications. But their continued utility and expansion into other more complex environments depend upon the rate of technology development that will support neural network systems implementations. In addition, their use also depends upon their ability to be integrated into the processing environments of more complicated systems. Network management, and in particular signal processing and machine vision applications, are particularly suited for neural network analyses and resolution.

Larry L. Ball, Ph.D.
Vancouver, British Columbia
September, 1993

Network Management
with Smart Systems

Part

1

Network Management Potential with Smart Systems Technology

Part 1 Highlights:

The first five chapters will discuss the technologies integral to smart systems' impacts upon network management. They are both hardware-and-software-based, being smart sensors, data fusion algorithms, and decision support tools. Some of these technologies are not new, but have become more useful as a result of recent progress in chip packaging, speed and processing techniques associated with analyzing the data made available.

The end results of employing these methodologies are improved decision making at all levels of the network management system. Chapter 1 provides an overall survey of the technologies available, while the next four chapters discuss each of the enabling technologies in more detail. Chapter 2 discusses the functional requirements of the smart systems issues, an important topic for the generation of a quality product(s). Chapters 3, 4, and 5 discuss the major topics as identified above.

Chapter

1

Smart Systems: Technology Survey

Chapter Highlights:

Smart systems technologies as applied to network management encompass three main topics. These topics are data sensors, data fusion, and decision support. Each of these may involve similar technological applications; however, each is unique in its functions, techniques, and methodologies that are utilized to support network management. For example, the sensors are usually located with the system components, while the fusion and decision support are usually performed at other locales. The whole area of smart systems technology is an outgrowth of military systems, in particular the command, control, communication and intelligence (C^3I) technology development work that has occurred over the past 50 years. In addition, the evolution of technology itself has greatly influenced the advancement of smart systems work applicable to network management and has currently resulted in capabilities not possible just a few years ago. This chapter surveys the present status of technology related to sensors, data fusion, and decision support.

1.1 Perspectives on Smart Systems

The whole thrust of technological innovation from the industrial revolution onward has been to extend and improve upon the capacities and capabilities of the human operator. Intelligent systems, to be referred to here as *smart systems,* are merely another step in that logical extension of human abilities in the telecommunications arena. To be more exact, smart systems are aimed at the extension of our ability to reason and learn. In order to emulate the human abilities for reasoning, associating, discriminating, and interpreting, one must turn to the model from which the conclusions are drawn, i.e., the human brain. Many such references are made to the brain in several places in this book.

Smart systems are viewed as subsets of the total apparatus of the system that oversee some purposeful mission, in this case the telecommunications network operation and support. These systems have the capability to draw conclusions and implement or recommend decisions based upon the inference of and reference to facts available to those systems. They, of necessity, must be able to generate, transmit, receive, and interpret communications between themselves and the controlling elements of the system, these controlling elements, many times, being human. In addition, they must be capable of some form of reasoning or logic process that allows them to reach their inferences or their recommended reference points. And, finally, smart systems must have access to real-time or external information sources and the reasoning framework, called by such names as rule base, frame base, etc., necessary to act upon these external data sources.

The importance of smart systems with regard to telecommunications technology developments, and in particular network management, cannot be overemphasized. In today's competitive environments, coupled with differing communications options available to the buyer, such as wireless, broadband, etc., the availability of the telecommunications platforms and their services is critical for ongoing customer satisfaction. This availability can only be realized through effective network management systems implemented using smart systems tools.

1.1.1   Characteristics of Smart Systems

Smart systems can be categorized to include a variety of expert and neural systems applications, usually implemented in software, ranging from maintenance advisors, to decision support modules, to pattern understanding, and sometimes including natural language interfaces for the human operators. The characteristics of a processing entity, that qualifies it as being an intelligent, or smart, system are summarized below. [3] Network management systems, for which smart capabilities are contemplated, should be examined for compatibility with these characteristics. The subsequent design of such system features must then be integrated with the target network management systems, and made consistent with the characteristics identified below.

(1) Coherent Knowledge

The system's knowledge should be coherent, i.e., it should be capable of being shared among all the modules within the system, be commonly available to these modules, and adhere to a common formatting structure, taxonomy, and lexicon. In conventional processing programs, information is segmented and parsed so as to be available to only one or a few processing modules. Distributed database systems, in a far-flung and complex network management system, provide this coherent knowledge reservoir by providing a consistency and accessibility of data.

(2) Integrated Access

All data available to the system should be accessible to all the intelligent modules regardless of the potential variety of data structures and access methods used in the individual database systems servicing the system. Conventional programs are usually constrained to interface with one database system because the interface driver software is designed for only one database system. Integrated access provides the capability to reformat and translate data files from one system environment to another. Thus, the subnetwork managers must be supported with data translators, so that access to needed data can be had. Integrated access also requires that a security system be established so as to provide appropriate levels of access privileges.

(3) Decision Analysis

To some greater or lesser extent, the reasoning and understanding used by the system should be made available to the user on demand through the use of informal-style questions that the user can pose to the system for resolution, or as part of a more lengthy dialogue with the system. In conventional software, the algorithms used by the program are seldom made known to the user during module execution. At a minimum, these algorithms should be made available to the user, the logic used should be published, on demand, to the user, and decision criteria should be made available as required. This facility enhances the network manager and gives its decisions or recommendations credibility.

(4)   Decision Execution

The execution of decisions either reached by or supported through smart systems, is a more far-reaching proposition that will entail further research and incremental development work in this area. However, as this time arrives, the affects of intelligent systems can be hypothesized. Intelligent submodules (smart sensors) should, eventually, be able to initiate appropriate actions on their own as conditions warrant. Execution of these modules depends upon external stimuli as opposed to user initiation. These smart sensors can also be integrated into a *data fusion* module so that filtering and data integration can make decision support more accurate and timely. The topic of *data fusion* will be discussed in more detail in the next section.

(5)   Acquired Knowledge

The intelligent modules should be able to extend the expertise of the overall system and to identify inconsistencies. Because these programs are dealing with a view of the entire system, they are in a position to determine the limits and extent of the system's capabilities as well as identify analytical inconsistencies. Knowledge is currently acquired through information input by way of rules, or through references to desired results. In such circumstances, the rules involved are updated from outside sources as well. Research and some limited development efforts to date have attempted to train systems to acquire and use acquired knowledge to reach new conclusions. These systems are known as neural networks, and are based upon a human-interpreted emulation of the operation of the vertebrate brain. The acquisition and use of knowledge in network management, such as that provided through neural networks, makes the operation more stable, and system reports more believable and accurate.

(6)   Non-Domain-Specific Problem Resolution

Further down the product development road yet is the ability for the smart system to solve problems outside the realm of the intended application. This is known as non-domain-specific problem solving. The intelligent system should be capable of solving problems not anticipated by the developer. Sometimes the smart system leads to a point of departure not anticipated by the developer by identifying a condition that requires further investigation. These requirements are best handled through a system capable of learning on its own, such as neural networks. The technologies of network management are constantly crossing into areas not anticipated by the developers. This tool, if properly implemented, could be a great time saver and problem solver for the system administrator.

1.1.2   Data Sorting and Characterizations

Here we will begin to survey and explore the various elements of the network management system that can be implemented with smart features. To begin with, there are several areas in the network system into which smart systems technology can begin to be infused, as a result of architectural design. These areas include front-end monitoring, and *data fusion*, which is closely aligned with the overall data collection process. *Data fusion* is the physical and logical integration of temporally and spatially dispersed reports. Data fusion, although not always explicitly stated in discussions on network management, is also an integral task in any network management system even though it may not be implemented to any detailed level.

The first area is the front-end monitoring of system components. Sensors monitor, evaluate, and report on component performance, and status. Sensors, as a general category of performance monitoring, have been around since the advent of electronics. Up until recent times, they have been relegated to the position of reporting on such problems as framing errors, transmission errors, and the like. Since the system sensors, as we will characterize them, are operating as elements within the network components in locations remote from the monitoring and control points, we would like for them to be somewhat autonomous in their roles and actions. The various functioning components and elements of the telecommunications network are usually the sources of status information about the network that is needed for the assessment of the total system. Each is a sensor of sorts that collects and reports on the status of its environment.

The second area into which smart systems technology can be infused into the network is that point at which the sensor data are brought together, and typed, sorted, and categorized. This is the data fusion function, and comes in two varieties, these being direct fusion and indirect fusion. Direct fusion is an integration of sensor data in a local environment, e.g., among the various equipments within a communications node. Here certain problems of distance, communications relay, and so forth, are minimized or negated due to the localization involved.

Indirect fusion is the integration of widely dispersed sensor data collected in, and through, many nodes and transmission paths. Information flowing into these areas must be continually evaluated both individually and in relation to other data in order to track the status and performance of the system. It is interesting to note that the very act of data integration has an impact upon the conclusions of system and subsystem health, and when added to the fact that synergistic data can be derived from multiple inputs and/or confirmations, the whole issue of data fusion becomes even more powerful. Unfortunately, data fusion requires additional effort to implement and even more effort to take full advantage of it. Thus, like everything else, there is a trade-off to be considered when considering its relative advantages versus its disadvantages.

There are six main tasks associated with data fusion. [17] Association is especially critical, since it is the point at which reports from seemingly divergent areas, but concerned with the same objects, are matched. Correlation is then invoked to determine the extent to which these initially matched reports are actually related. These data fusion tasks are:

(1)   Collection

This task is performed by the sensors at the collection sites specifically, or as part of the relay process of data from the collection sites to the fusion site(s). The collection process is critical for the data fusion activity, because the data reports have to be separable and definable for later correlation and collation. The number of incoming reports can be staggering, so a network management system must be provided that meets these data rate needs. The collection task can help mitigate the number of incoming reports by examining reports for redundancies, duplications and repeats, and eliminating such reports.

(2)   Alignment

This is the point at which filtering occurs. Filtering consists of clearing up ambiguity, as much as possible, about the individual incoming reports. Additionally, alignment means to assign credibility to individual reports, so that such reports can be given even more confidence as the data fusion process continues to the next steps.

(3)   Association

This task involves updating and composition of report states. The task of associating reports, either the same reports over multiple time periods, or different reports relating to the same circumstances is handled in this task. The object here is to match different reports with the same object or event, and to single out redundant and/or unnecessary reports, either from the same or different sources.

(4)   Correlation

This is the point at which ambiguity resolution occurs when the association process cannot resolve ambiguous reports. Correlation is of its nature a chance situation; thus, techniques suitable to the specific network management situations must be employed to minimize the probabilities of errors.

(5) Classification

This task is directed at the identification of a class of problems from which more specific elemental identification can be made later. The integration of several sensors is usually involved accompanied by some form of decision process to specify class I.D. and confidence level. In a network management environment, classification may involve categorizing reports into different levels of alarm, different levels of maintenance attention, different levels of operational alerts and so on. Thus, a multilevel, multicategory set of needs arise.

(6) Control

This task is involved in the allocation of direct sensor systems, their optimization, and their prioritization. Control encompasses a range of tasks and functions too vast to address here. In essence, the control subject involves the acts of delivering the right information to the right point at the right time. Decision support/executive information systems are all about the problem of control and its ongoing resolution.

Because of the increasing complexities of equipment, functions, and sizes of networks, the various smart systems technologies to be addressed here are critical to the effective control of telecommunications networks.

Data fusion is a more complex subject than it may appear. The network data involved are many times from vastly different environments and from different vendor equipment arrays. The formats vary from each other and the correct interpretation of data contents can be very different. Data fusion is also the point in the data collection and analysis functions at which the confidence of incoming status data is confirmed or calculated. As a matter of fact, the major reason for implementing data fusion is to verify network status indicators, and to assess confidence in the ensemble of incoming data elements.

Data fission is the opposite of data fusion. Data fission is the process of finding an object of interest in an array of previously collected objects. Data fusion involves the act of sorting individual objects, locating, then labeling them for future fission operations at which point they are split out for usage. Clearly these two processes are mutually dependent, because the efficacy of the fused data makes searching very dependent upon the time element which in turn determines the usefulness of the total process.

The last area of smart systems infusion into the network management scheme is the point of decision making. This point is often referred to, in the defense sector, as decision support, and, in the commercial sector, as the executive information support function. Decision support involves the use of categorized and sorted data from the data fusion function, coupled with decision criteria, frames of reference, decision criteria, and the decision maker's preferences, to arrive at a rational course of action.

Each of these topics will be briefly addressed and associated with each other before a more thorough examination is made in the ensuing chapters. Shown in Figure 1-1 is a concept of how these technologies fit within the overall scheme of the network management approach. The smart system concept is broken down into three main segments, namely, sensors, data fusion and decision support. The primary operational segments of the communications system, its data converters and transmitters, its switches and relays, and its monitoring and control units are each roughly equated to each of the smart systems segments, e.g., sensors are associated with the data converters and transmitters, and so forth.

The selection of an architecture that provides the system coverage, responsiveness, and diagnosis capabilities is always difficult, partly because no two telecommunications systems are the same, and partly because these systems lack homogeneity. The typical approach has been to integrate individual network management subsystems using an additional network manager that recognizes the attributes and characteristics of each of those subsystems with which it interfaces. Another approach is to bypass the subsystem network managers, and to route each of the equipment elements to a single network manager. Yet another approach has been to utilize the existing performance characteristics of the subsystems and to infer the maintenance and performance of each equipment on the basis of these characteristics. Smart systems greatly enhance this latter approach.

The intelligent features of each of these segments is infused into these segments via the main avenues defined, i.e., sensors, fusion, and decision support. Thus, the sensors associated with the data converters and transmitters are designed and implemented so as to incorporate Artificial Intelligence (AI) features and autonomous operational capabilities. Similarly, data fusion is associated with the switching and relay elements of the typical system, and its features are designed to incorporate expert systems capabilities. Likewise, the decision support function is akin to the monitoring and control elements, where their functions are incorporated into the design of decision support capabilities that reflect AI features.

The overall network management scheme is to monitor operating units for status and health, and to take action when and where necessary to continue the communications traffic flow. From such collected data, there is an assumption that all other needed information can be derived. This information is periodically or continuously relayed to a point where it is analyzed or in some way organized. From there it is evaluated and decisions are made based upon the logic used for such decisions, the criteria established for decisions, and the efficacy (appropriateness) and accuracy of the data categorization.

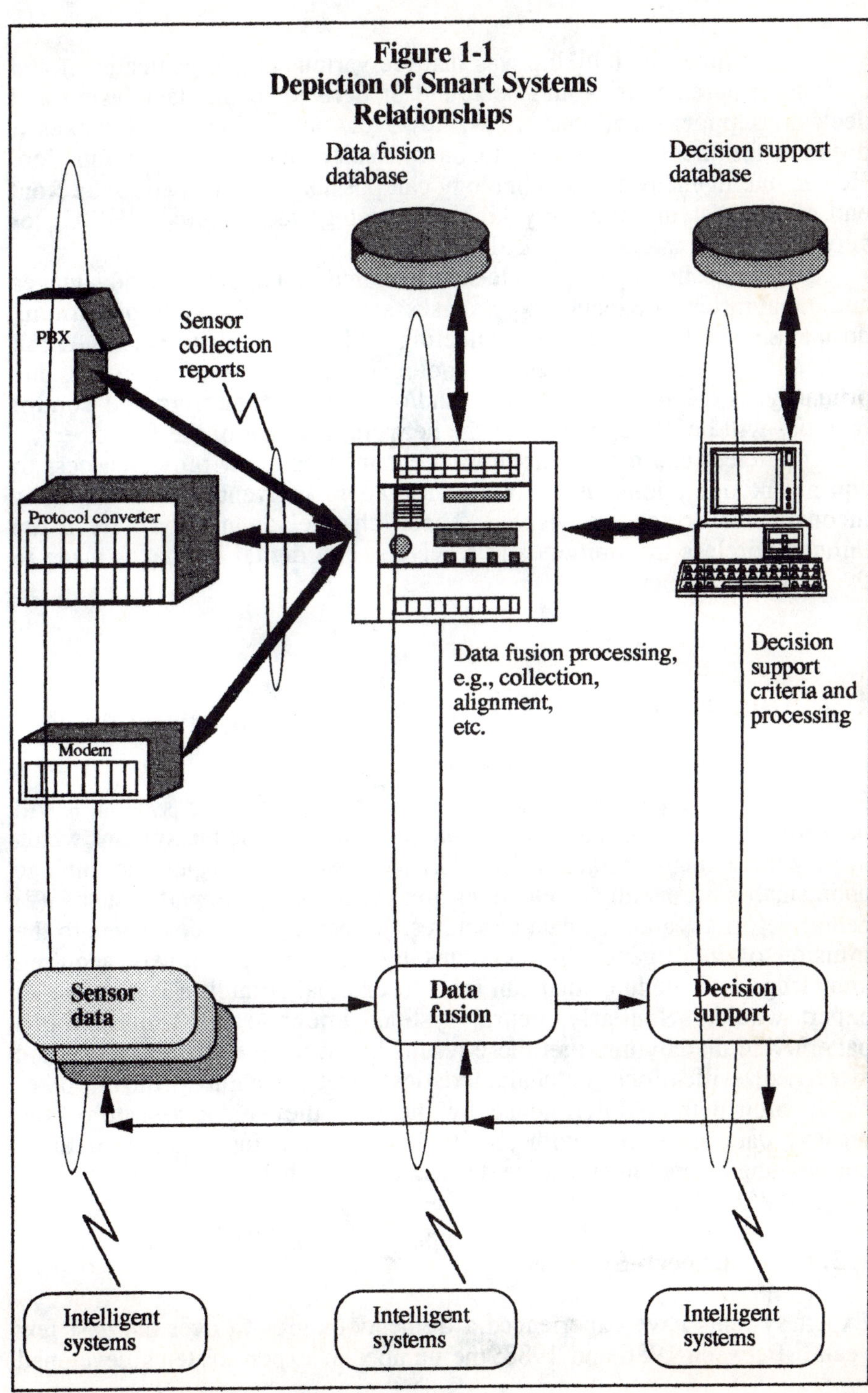

Figure 1-1
Depiction of Smart Systems
Relationships
Data fusion
database
Decision support
database
PBX
Sensor
collection
reports
Protocol converter
Modem
Data fusion processing,
e.g., collection,
alignment,
etc.
Decision
support
criteria and
processing
Sensor
data
Data
fusion
Decision
support
Intelligent
systems
Intelligent
systems
Intelligent
systems

Until now, the thinking was that the various data organizational and evaluation parameters would be found or derived in the data fusion and decision support functions. Now, however, processing power makes it possible for these capabilities to be offloaded into the sensor function. Hence, intelligent systems technology can possibly be imparted to the front end of the system using very high scale integrated circuits (VHSIC), or very large scale integrated (VLSI) circuits.

These technologies provide sizing, memory, and software advantages and capabilities to execute expert tasks at high speeds while performing normal sensor tasks, such as collecting and reporting the data. At the backend or control point, these technologies can also be of some use, due primarily to the fact that wireless and/or mobile monitoring and control facilities will be of importance to the network manager of the future.

Wireless and mobile monitoring channels can now provide access to equipment operations in environments where conventional transport is inconvenient or too expensive. These channel connections are made through wireless transmitters and receivers, and digital interfaces made to the sensor components.

1.2 Smart Systems Technologies

Before beginning a review of the basic features of any smart system, it will be instructive to delve into the technologies that provide the systems which display these smart characteristics. The principal technologies that impinge upon smart systems implementations are: (1) rule-based expert systems, (2) neural networks, and, (3) data structures. Expert systems contribute to the infusion of intelligence into systems by encasing previously acquired knowledge into modules that can react to external stimuli in a way that an expert would. Similarly, neural systems integrate the stimuli into a pattern, and thereby interpret these results based upon their total integrated affects, classifications, or characteristics. Data structures, on the other hand, contribute to intelligence by the way they organize, store, and retrieve data associated with the intelligence functions. Each of these contributing technologies will be discussed briefly below.

1.2.1   Expert Systems

Expert systems have experienced a tremendous growth over the past few years. Between 1986 and 1988, the number of expert systems developed and used in commercial or government applications in the United States increased 215 percent from 475 to 1025. [14] Between 1988 and 1990 the number of systems available jumped again by another 147 percent to 1510. Even today, most expert systems development and delivery are executed on PCs using development tools costing less than $1000. Expert systems support practically all facets of network management, such as maintenance, planning, and operations. These issues will be addressed in more detail later in this chapter, and given individual treatment in later chapters of this book.

1.2.2   Neural Networks

Neural networks are an attempt to model and simulate the neural architecture and functioning of the brain. In particular, such models are an attempt to emulate the processing and interpretation of the various sensory modalities of the mammalian central nervous system. This seemingly huge task is accomplished through the use of simplified models of neural processing, as conducted by neurons (brain cells) that are arranged in interconnecting patterns. These patterns are, in turn, exercised by external stimuli to yield the correct responses to problems presented to them.

Neural network models can be developed on conventional computers from simple PCs to parallel processors, and supercomputers using special software developed for such purposes. [8] In addition neurochips have been developed embodying the capabilities to perform neural network simulations. Finally, application specific integrated circuits (ASICs) have been developed for individual situations that have the ability to execute and analyze a specific neural net problem.

What is a neural net problem? A neural net problem is one for which there are no set of formal or articulated rules. A neural network problem is, therefore, one that is solved by trial and error using a training paradigm that is exercised until it yields the right set of outputs given the inputs presented. Thus, because neural nets are trained rather than programmed, they are especially suited for classification and pattern recognition problems. Examples of applications to which neural networks could be applied include diagnostics, machine vision, modeling and theory development, pattern recognition and classification, signal processing, speech recognition and classification, and robotics.

For network management systems and applications, neural networks have widespread possibilities. Neural network inputs that might be equated to different features of an operational network can be routed and applied to the first layer of a typical neural network. The subsequent processing can generate a pattern, the output of which can be used for reference, or to modify the neural network so that it attains a reference point. This pattern now becomes one of the comparison elements, against which pattern alterations can be judged. This paradigm is well illustrated in Chapter 10.

1.2.3   Data Structures and File Management

While not usually thought of as an essential ingredient of AI, data structures and file management facilities contribute greatly to the infusion of intelligent features into systems because they provide the facilities to speedily locate or properly associate data items of interest with each other, and with the subject.

A good case in point is the subject of list processing. An AI program known as LISP, short for list processing, was developed to facilitate the association of objects and their attributes with each other. List processing is a file structure facility where lists and taxonomies of different categories of items are organized for later recall, sorting, and reorganization, as directed by the LISP commands.

Cited below are some of the most important features of data structures and file management technology. These features, even though complex in some cases, fall into a few basic categories, these being searching, sorting, and updating.

Data Searching

Data searching is the process of accessing a particular piece of information. Many times this search process will entail the use of a key value that is best defined as the argument of the search. [11]   Search techniques can be identified as being either linear or nonlinear. There are several varieties of linear search techniques that may be applied to finding one item out of many. For the most part these linear search techniques require on-the-order-of (O) N searches (worst case) or $N/2$ (as an average) searches to find the item of choice, where N is the number of total items from which to choose.   On-the-order-of is roughly analogous to rough order of magnitude.

Nonlinear approaches can have fewer steps than the $O(N/2)$ steps required by linear techniques. One of these nonlinear techniques is the binary search technique that has the smallest number of required steps. This approach requires $O(\log_2 N)$, and can be used to search either records or smaller entities, such as data fields.

Another series of search techniques is involved in file management where the data are arranged in records. In such situations, searching can be accomplished by the use of addressing mechanisms that utilize keys [11]. The three principal methods of searching using these methods are direct mapping, directory lookup, and calculation. All of these can be thought of as a lock and key combination, where the key is the name of the item of interest and the lock is the manipulation that allows the address of the item of interest to be identified. This concept is illustrated in Figure 1-2.

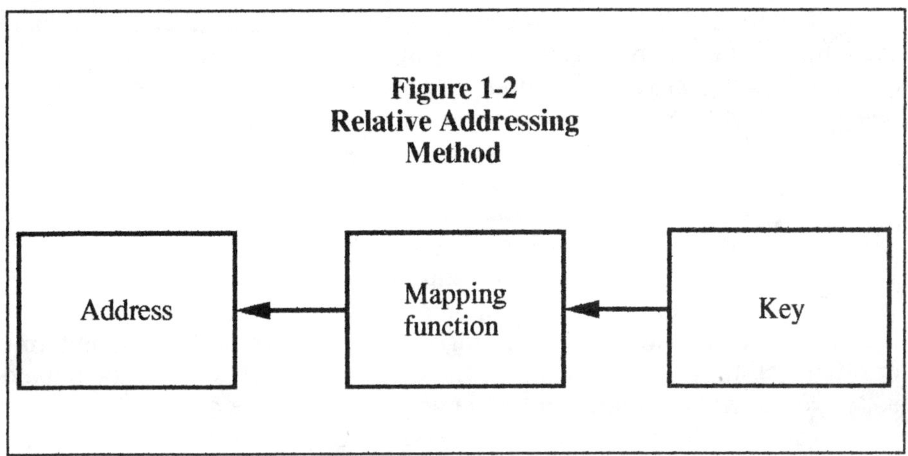

Direct mapping is a search technique in which the key of the data item to be located is, in fact, the address of that data item. In such a case, there is no mapping or translation of the key per se. For directory lookup, the key will point to some location in a separate directory where the address will be found somewhat like the use of a telephone directory. The final type of search technique is also called "hashing," or calculation, after the meal that contains many ingredients. The hash technique takes the key and transforms it into the address using any number of different algorithms. The algorithms do not have to make logical sense if they accomplish the required task. One of the hash techniques used is called division remainder. In this technique, the key is divided by some appropriate number, the remainder of which becomes the relative address of the record or data item sought.

Sorting

Sorting is important for database systems involved in network management because it directly supports the storage of parameter values associated with factual data, and the searching techniques that perform access operations to retrieve that stored data. The normal method of storing data is to assign a storage entity, or record, for each parameter of interest. These records may be either sorted or unsorted as they are stored in the database system. If the records are sorted, then searching (retrieval) is greatly speeded up. Sorting is sometimes categorized as either being internal or external, depending upon whether the sort routine can be performed in existing memory or whether auxiliary memory is required for the operation. External sorts depend for speed on two major features of the system, one being the read and write times of the hardware and software support system, and the other being the speed of the sort algorithm itself. There are families of sort algorithms associated with data arrangement, these being selection sorts, insertion sorts, and exchange sorts.

The selection sort consists of surveying a set of keys and selecting each key in turn in either ascending or descending value. Such a sort is said to require $O(N^2)$ comparisons for completion. The insertion sort is a technique in which an unsorted list is again subjected to scanning, and each key is taken from the unsorted list and placed into a sorted list such that each key is inserted into its proper position in the sorted list as it is selected. The class known as exchange sorts offers the best results in terms of efficiency in the sorting process and consists of such sorts as the bubble sort, quick sort, heap sort, and tournament sort, to name a few.

The tournament sort is probably the most popular of all sorts today. It was first described in 1956 by E. H. Friend, a noted mathematician. [11] It is used today by many database vendors in their file-sorting software packages, and its methodology is illustrated in Figure 1-3. The tournament sort operates much as the tournament elimination process in sports would operate except that we can sort in either a descending or ascending order.

For an ascending ordered list, the process begins with a raw list of values and the first four values, divided into two pairs, are initially selected for comparison. These two pairs of values are compared for minimum values in each pair. The smaller of each comparison is selected and again compared to the "winner" of the first comparison. The smaller of this comparison is placed in a holding position.

The value ultimately selected is replaced by the next unsorted key in the raw list, and the process is repeated. This activity continues until all raw keys are processed. At some point in the sort process a new raw key is selected that has a value that is smaller than the last key output to the processed list. In this event, the new raw key is placed in the sort tree but eliminated from consideration unless and until all the branches of the tree are filled with keys smaller than the last key output to the processed list.

At this point the process is repeated to generate a new list that has ascending ordered keys. There are now two lists, one being the processed list and the other being the keys remaining in the comparison slots. These two lists can now be compared and merged as previously discussed.

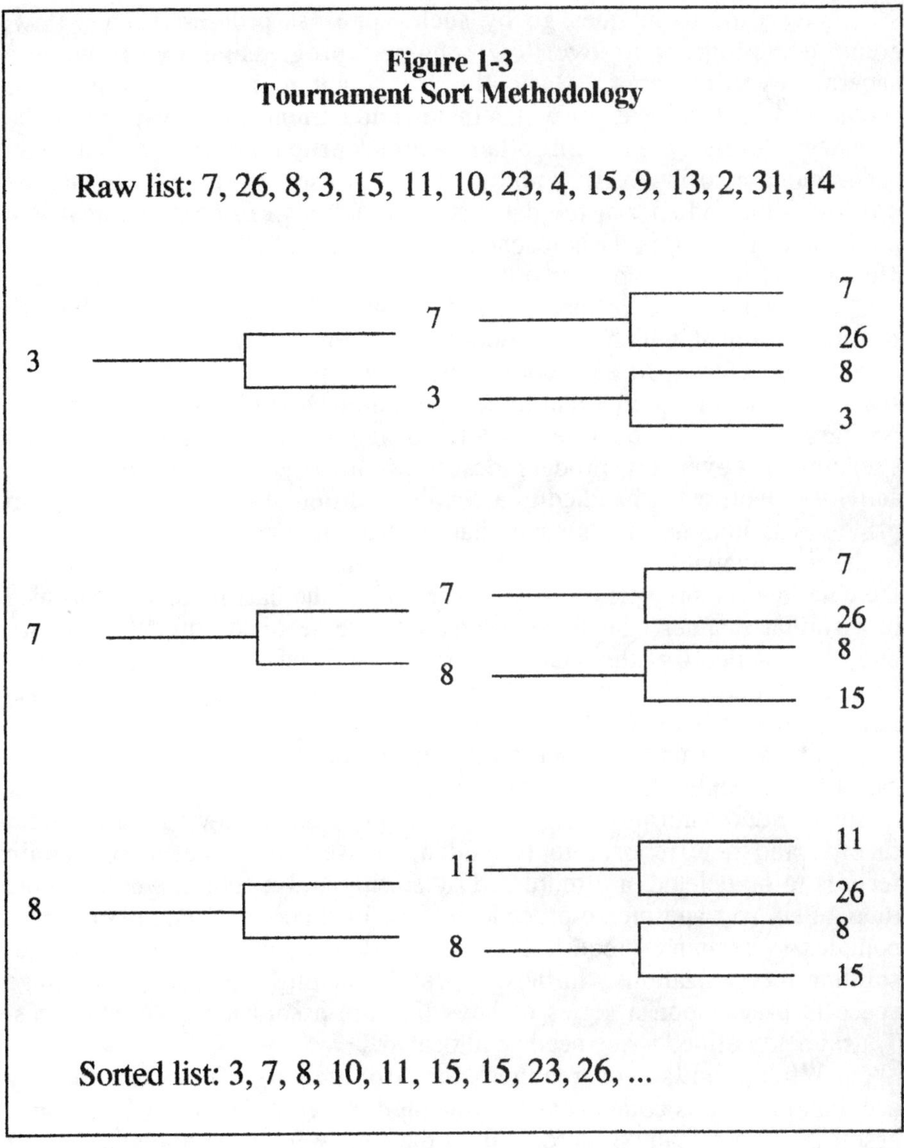

Updating

Updating is a process of periodically reorganizing and reconstituting information sets. It consists of two phases, one being the handling of overflows and the other the actual update itself. Database systems utilize several techniques for data updating, all of which are a function of the overflow techniques used, and these are explained below.

Overflow techniques go by such names as progressive overflow, count-key progressive overflow, chained progressive overflow, and separate overflow area. Basically, the problem in data storage is to accommodate data items for which there is no current space available in the primary storage area. In other words, primary storage can only accommodate so much data, and after this is used up, temporary storage must be allocated so that the database transactions can continue unabated until some stopping point is reached, at which time the data records can be reorganized into their proper order.

Storage can be defined according to specified addresses within the medium, each of which may contain room for one or more data items, sometimes referred to as records. If, at any point, there is not enough space for a data item, the item to be inserted must be placed in an overflow position. How this position is defined and what the algorithm is that executes the overflow process describes the first phase of the update activity. Overflow is handled in a couple of different ways. One way is to place a data item that finds itself shut out from its primary storage position in the next available storage location that is free. Another way is to place the data item in an overflow area. Retrieval of the data items is performed in a similar manner. Successive locations are scanned, after the primary one has revealed that the data item of interest is missing, until the item of choice is found. Alternatively, the overflow area is searched until the data item of interest is found.

The actual update process may be solved in different ways. In the case of a sequential file, one in which records are stored one after the other in some ordered form, the update may require the following: all current records and new records to be added, as well as requests for certain records to be deleted, are reordered by sorting and merging, similar to the tournament sort and merge procedures described above. Updating of more complex systems may require elaborate methods, such as various storage-splitting reorganizations. In these types of updating processes, the storage space is based upon a series of keys that are associated with addresses. Thus, no predefined space need be allocated.

When limits are reached for storage, overflow temporarily accommodates this component until the update can be performed. In some cases, the update can be performed on the fly. What this entails is that the number of keys at each storage location, which have reached their maximum, are physically split such that two groups of keys are made from one and the groups are linked by the use of pointers. This is illustrated in Figure 1-4. Techniques such as this one in various forms, are used in balanced tree, B tree, and other database design approaches.

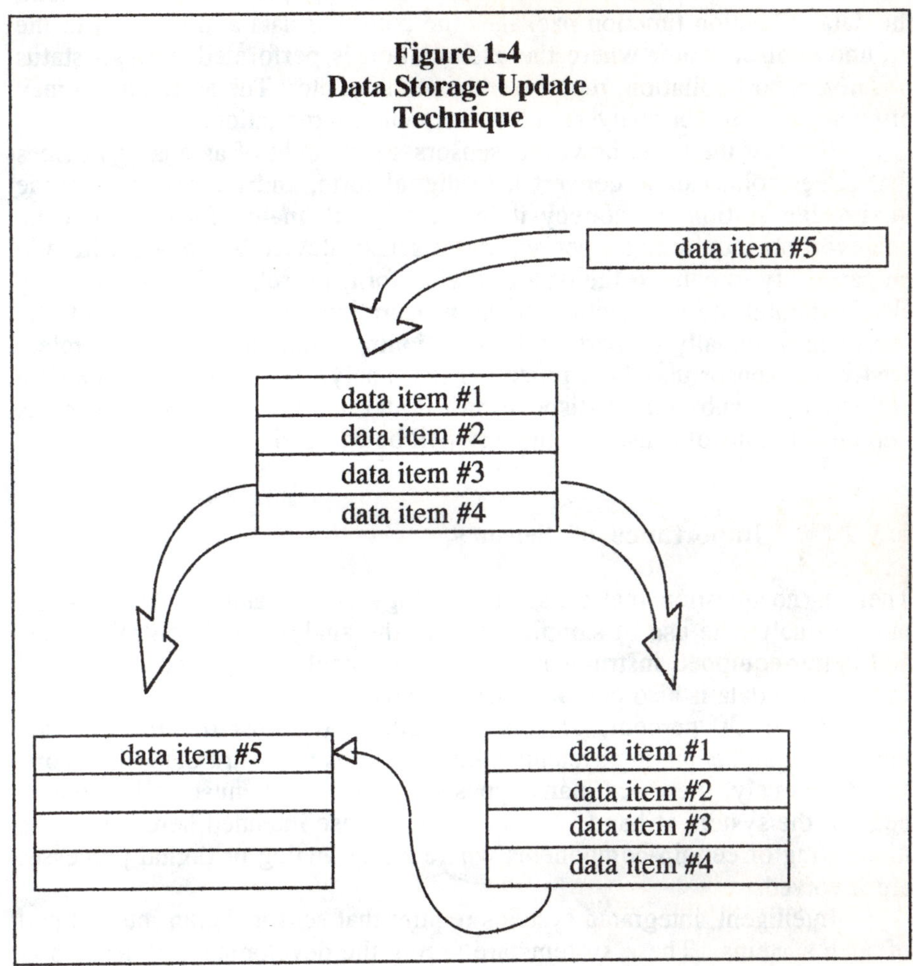

1.3 Sensor Technology

1.3.1   Definitions

A sensor can be roughly described as any reportable data source, but more often is described as any device that gathers data. [6] A data source could contain performance, configuration, or accounting information that might be correlated with other time critical data to arrive at the system status. Sensors consist of two basic ingredients, namely the data collection function and the data assimilation function. In a satellite operations environment, the data collection function packages the collected data and relays it to the ground control center where the assimilation is performed through status updates, report collation, report trend assembly, etc. The assimilation may often suggest further analysis and subsequent interpretation.

Most of the time, however, sensors are thought of as analog devices that either collect data, convert it to digital form, and transmit it to some monitoring station, or convey it in its original analog form. There is, however, no requirement, per se, that a sensor device be analog with add on capability to convert the data to digital form for relay. The sensor may also be digital in its collection mode of operation. As a matter of fact, the sensor may actually be part of the functioning component being sampled. Lastly, the sensor may be a processor that analyzes and evaluates the data collected for subsequent disposition. Later in this section the various implementations of sensor technology will be categorized.

1.3.2   Importance of Sensors

There is no question that sensor technology is increasing in importance, mostly due to the use of sampled data in the analysis and control of new and better-equipped instruments for human benefit. This increased need for sampled data is also partially due to its use in expert systems. By 1990, for instance, 30 percent, or about $2000, of the cost of all electronic monitoring and actuating systems built into autos was attributed to sensors. [1] Obviously, there are many types of sensors that must be brought to bear on the system at hand. *Sensors* in the sense intended here means the monitoring of certain components where either analog or digital processes are involved.

Intelligent, integrated systems require that sensors be an integral part of such systems. These systems are seeing the development of sensors as intelligent peripherals to the main processing system that can be programmed and accessed by more than one system processor. There are four phases to the evolution of the integrated sensor, these being the expansion of the micro-structured types and varieties of sensors available, the expansion of the functionality of the sensor, e.g., the filtering and compensation; self-diagnosis and calibration capabilities of the sensors, and finally the intelligent peripheral state of the sensor that makes it a reliable and valid partner in the system operation. [1]

The types of sensors available to the designer are discussed below. While the major discriminate is analog versus digital, the two can easily be integrated, one within the other, as for instance, when digital status information is buried within the analog frame of a TV picture.

1.3.3   Types of Sensors

Sensor systems must be thought of as integral to the realization of complete automated systems both currently under development and slated for development in the future. Digital, bus-capable sensor interfaces are one of the solutions foreseen for future applications of sensor technology that will support integrated sensor and operational functions. These interfaces can best be accomplished through the definition of an open sensor interface that offers the flexibility to exchange the equipment of different manufacturers. [7]

Multipoint sensors are important to the effective management and tracking of connectivity within a system. Figure 1-5 portrays a configuration consistent with the Open Systems Interconnect (OSI) approach to telecommunications organization for a digital bus sensor technology. [13]

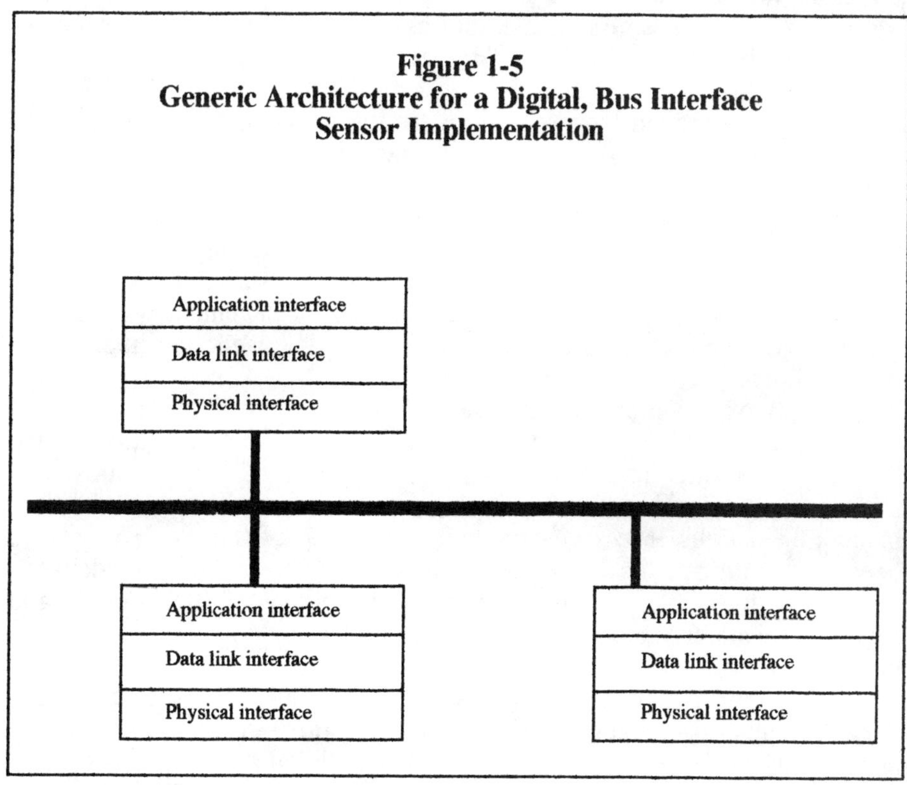

Figure 1-5
**Generic Architecture for a Digital, Bus Interface
Sensor Implementation**

In this figure the physical or network interface provides the connection between the network cabling and electronics and the device attached to that system. The data link interface provides the data structural connection between the two entities, while the application interface provides the informational exchange between the network and device.

Digital Sensors

Sensing in digital systems is normally accomplished by one of several methods, one being to convert analog signals to digital format and then incorporate that data into messages that are transmitted between the affected equipment and the network management system. Another method is to monitor the direct digital sequences generated by the target equipment and to incorporate that information into status messages for transmission to the network management system. This all assumes, of course, that the transmission system itself, which carries the status messages, is intact. If it is not, the digital sensor paradigm fails because the transport medium itself is missing. The transmission medium over which the status messages as well as the operational messages travel is clearly the weak link in the overall process.

One way around the problem of having to use a common transport medium is to make use of a wireless or a radiofrequency (RF) link between sensor and monitoring point. Thus, the sensor, no matter how implemented, can provide reliable backup communications for monitoring and control purposes via a separate RF link that interfaces with the sensor unit. A second and more common method of assuring reliable status communications between equipment and network management systems is to provide redundant circuits, one of which is used for the status transmissions.

Short of complete failure, the transmission medium provides the communications link(s) between the sensors and their reporting points. Digital sensor interfaces are implemented via one of two methods, i.e., message passing and rendezvous techniques between target units and their monitoring stations. Message passing is the process of conveying a message between a unit and its higher-level monitor. The message rendezvous process is the technique of conveying a message between unit and supervisor via some intermediary such as an electronic mail box or storage point.

The fact that digital sensors are operating on discrete data at irregular intervals makes it difficult to treat them in the same way as analog sensors. One of the ways to rationalize digital sensor data is to view digital sensor output as a continuum tracing the movement of events as if they were readings that had been subjected to an analog-to-digital (A/D) conversion. Another way is to view the digital output as a time series trend analysis. Both of these issues will be dealt with later in another chapter.

Analog Sensors

Analog sensors have been around for many years and are found in almost all large-scale processes, partly because the modeling and simulation of big operations has been almost impossible until the advent of high speed, and large-capacity memory computers, thus making the data collection needed for large scale systems design a function of direct measurements. Additionally, until more economical and inferential measurements could be made, analog sensors have been necessary and useful in the monitoring of operational processes that benefited from any such monitoring.

Except for very high rate analog signal inputs, inexpensive technology is available from many sources to convert analog signals to digital formats for better handling.

1.3.4 Implementation of Sensor Functions

Sensor messages are of two principal types, these being status and diagnostic. The status messages alert the control station to potential congestion problems, their locations within the system, and their nature. Diagnostic messages alert the control station of different types of failure within the network element. These two types of messages are the result either of data collection by the sensor within the unit itself, with subsequent formatting and conveyance to the control station, or of status and diagnostic messages generated by the unit itself with subsequent conveyance to the control station. Sensory functions acting internal to the equipment are a different concept that deserves some attention. Such a process could be accomplished by a software module residing within the equipment or by a hardware/firmware device interfacing with the control station across the input/output (I/O) port(s) of the equipment item.

One very pertinent question that might be asked is, "How can these sensors be implemented within a typical equipment environment?" Potentially, there are four ways that this can be done and each is discussed briefly below.

(1) The network element itself can act as the sensor by collecting and periodically reporting status information about its processes.

This can be manifested as a hardwired capability without direct computer control, or it can be implemented under software control integrated with the basic system. An example here might be a signal repeater. The repeater receives incoming signals, boosts their power levels, and repeats these boosted signals on the output side. Thus, the repeater is a means of relaying a signal at some minimum level. Its status can be determined by its continuing functional performance at specified levels of acceptance, and not require further monitoring.

(2)   A separate software sensor module can be implanted within the equipment software environment as a nonintegrated element of the processing capability.

This entity can continually monitor and periodically measure the telecommunications subsystem's readiness and its status. A top level functional view of such a module is presented in Figure 1-6. There are

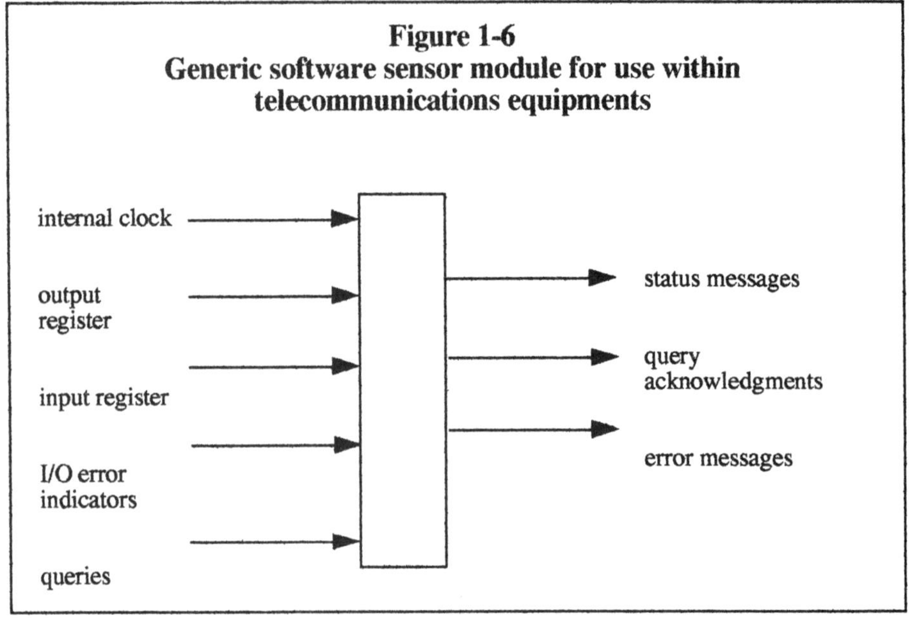

two principal modes of operation for the software sensor, one being the *error-trapping mode*, where an error initiates a message through the communications interface to the monitoring system, and the second being the *status report mode*, where periodic reports are sent to the monitoring system by the sensor module. Such a device could be implemented either with or without memory. A memory component would facilitate its ability to analyze data within the unit, to develop trends, and to report status at infrequent intervals.

(3)   A hardware device can be attached to the network element to monitor, collect, and transmit status and health information about the component to a central information control platform.

A very good illustration of this approach was suggested by IBM and is illustrated in Figure 1-7. [24] The cabling could be either a databus or local area network (LAN) configuration where the various devices to be monitored are controlled by a host computer, or some other device, such as a server or controller. Each sensor interface card consists of a specialized network attachment interface that varies with the communications protocol employed.   After that, the components are similar in that the same functions must be performed in order to sample, receive, judge, and format each piece of data examined.

Each sensor interface has its own hardware device acquisition subsection that would possibly be analog at the interface, an analog-to-digital conversion for those inputs that were analog, a controller that strobes the data reports into the random access memory (RAM) memory, its own read-only memory (ROM) instruction repository that determines the rates of data acquisition, and another interface to the specific network protocol in use, e.g., Ethernet.

(4)   A hardware device can be used to interface several network elements at a time so that sensory information can be handled at some centralized point.

This scenario involves the direction of sensory messages to one or a series of intermediate collection points where they are filtered and forwarded in an efficient and timely fashion.

1.3.5   Sensor Failure Modes

There are two major types of sensor failure which, not surprisingly, are referred to as hard failures and soft failures. [6] When a system element fails to respond totally to the requests of a sensor module, or when it fails to acknowledge a sensor module request, a hard failure has occurred. This is synonymous to a light switch or binary event.

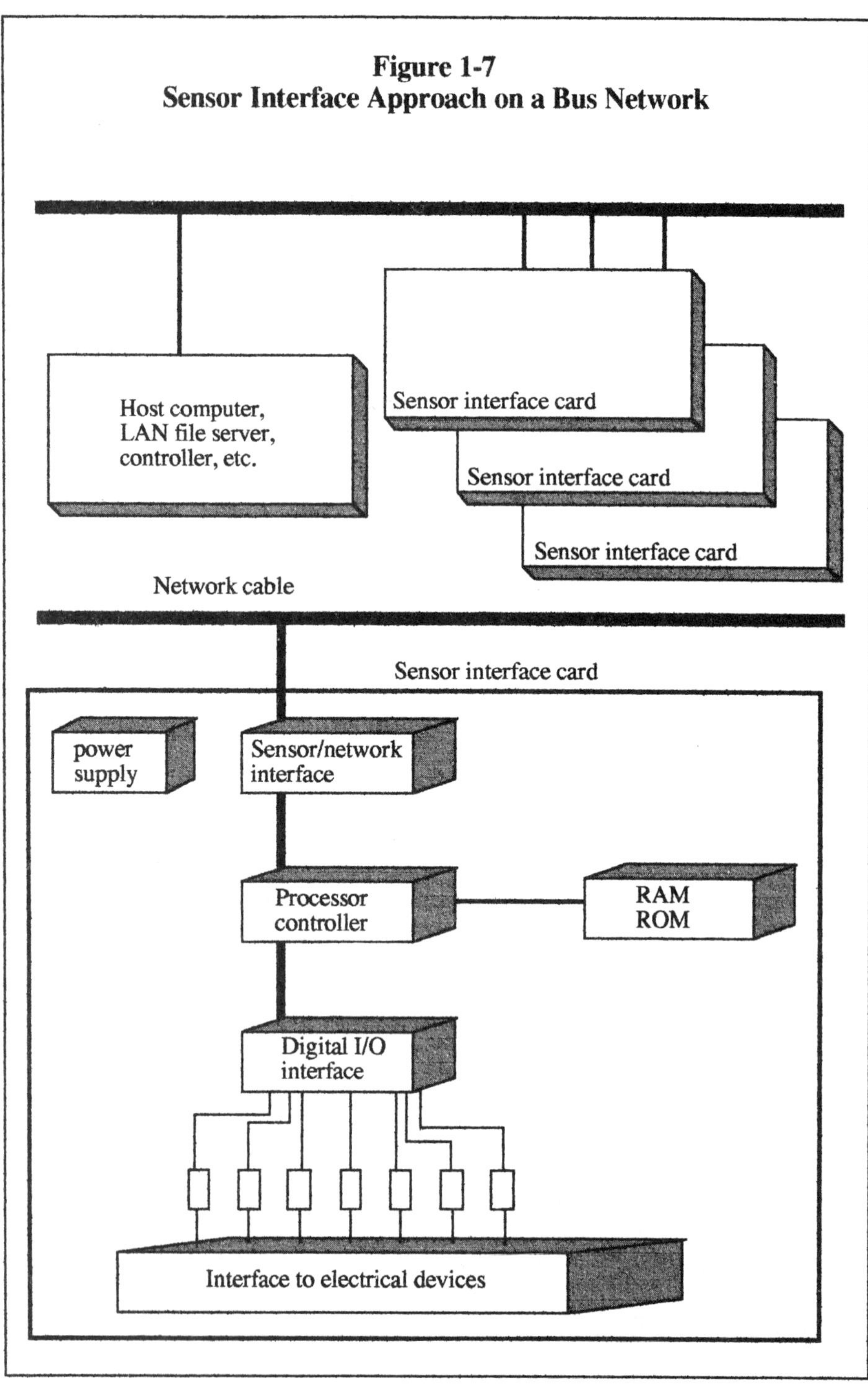

Figure 1-7
Sensor Interface Approach on a Bus Network
Host computer, LAN file server, controller, etc.
Sensor interface card
Sensor interface card
Sensor interface card
Network cable
Sensor interface card
power supply
Sensor/network interface
Processor controller
RAM ROM
Digital I/O interface
Interface to electrical devices

A soft failure encompasses that whole area of circumstances between total failure and complete operational availability. Soft failures occur when inappropriate or unfounded responses are sent from the equipment to the sensor. Such responses might result in garbled data, in undefined data sequences, or in illogical responses. Soft failures are also the hardest types of failures to resolve because their manifestations may be intermittent or based upon component failures that are difficult to trace because they still allow data flow between sensor and system element.

1.4 Data Fusion Technology

Data fusion is not a thing or technology per se, but rather it is a way of thinking about the data collection problem. [16] In this respect, it is an application-specific procedure that is implemented through a combination of software, hardware, and human-orchestrated procedures.

Data fusion can be thought of as the bringing together of two or more sensor reports from the same type of sensors or dissimilar sensors for the purpose of verifying and tracking conditions or events. The issue of data fusion stresses the importance of time, thereby provoking a series of reports closely spaced in time to speed up the analysis and association of events, and therefore the decision-making process. A series of reports from the same sensor may well generate the same answers to the questions at hand, but time and independent verification will be sacrificed, leading to decreased confidence or confusion at the decision point.

The utilization of several reports that cover multiple sources of interest is of much more value than a single report, simply because one reporting source does not convey as much meaning as do a group of such sources. This can easily be appreciated by considering what multiple source information imparts to the viewer.

1.4.1   The Data Fusion Model

As one might imagine, the data fusion problem is of major concern in a military environment and the basis for effective action in many engagement situations. The network management environment provides many similar parallels. There are several groups in the defense arena that have spent many years of effort to investigate and describe how data fusion can be implemented at the Department of Defense (DoD) level. The data fusion process has been described as having three levels of activity. Each of these levels is identified and briefly described below.

(1) Level 1

The position and identification of the fault or undesired condition is made here. Data fusion techniques are used to eliminate any ambiguity involved in resolving the point at issue. Isolating the point of a problem is a difficult task. The number and location of sensors directly affects the isolation of a problem down to some acceptable level of ambiguity.

(2) Level 2

The assessment of the situation is made at this point using techniques to determine the extent and magnitude of the technical problem. The situation assessment usually applies to the physical and functional attributes of the system. Specifically, the situation assessment evaluates the extent of the problem upon the elements of the system.

(3) Level 3

The threat or impact assessment is carried out here. This level has to do with the operational impacts of system or subsystem capabilities that may be compromised as a result of one or more component problems. In other words the object here is to determine what operational capabilities are impaired or stalled.

1.4.2   The Decision Problem

The object of data fusion is the proper classification of events detected. This process involves the tracking of several sensors at the same time, all of which may also observe the same, or nearly all of the same, set of events. There are several methods used for the solution of the fusion problem; however, the Bayesian method is many times cited as the archetype and best example for explaining such a problem. Figure 1-8 shows a schematized illustration of a scenario in which several connected elements provide input to a central control point. At this juncture, the data fusion activity occurs such that different possibilities are evaluated and acted upon. The data fusion process actually helps us to get a jump on the situation assessment that encompasses the operational impacts and decision support function. By using templates and rule-based approaches, situational impacts can be narrowed by the data fusion performed at this point. If the situation can be reduced to some short list of possibilities, the decision analysis becomes an easier task.

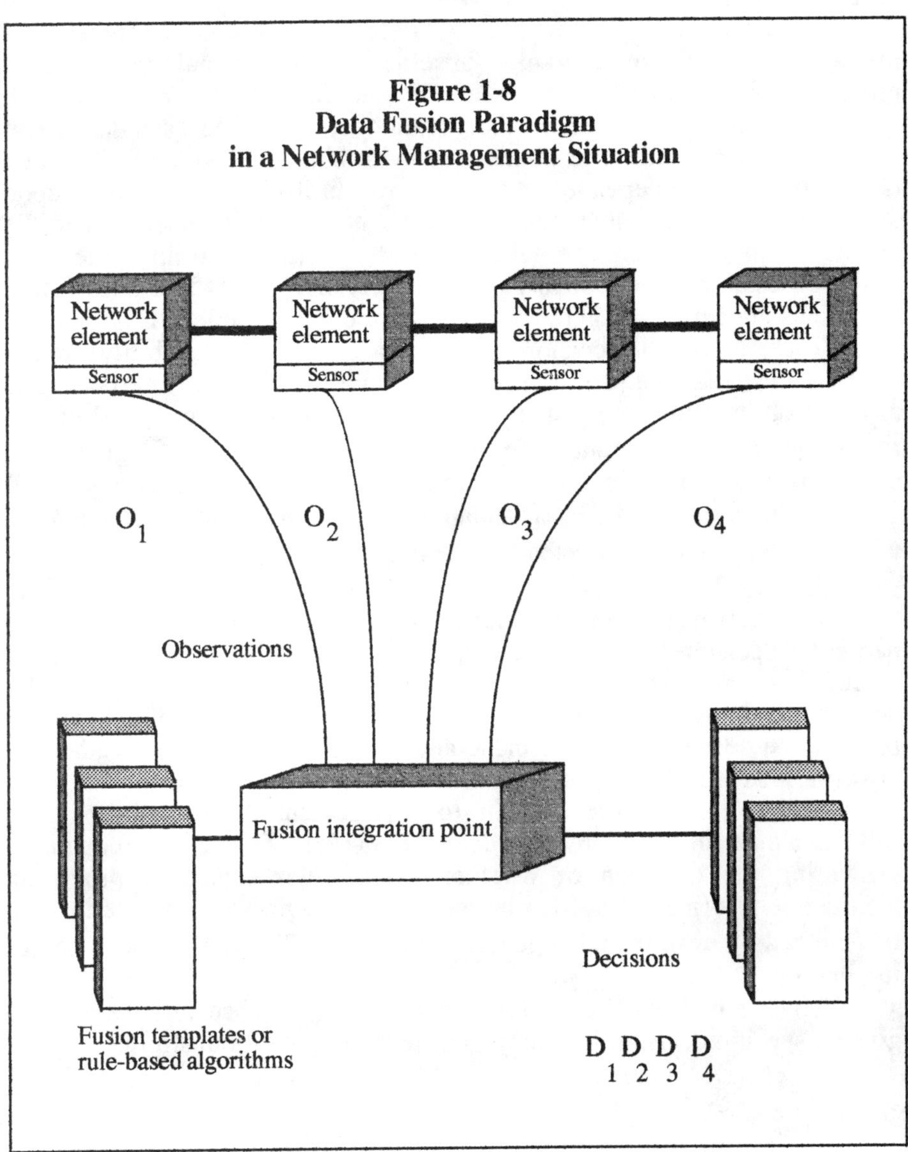

The matrix shown in Table 1-1 is a representation of the possible observations (O) and decisions (D) dealing with some hypothetical network operations situation. Bayesian statistics allows us to assign probabilities to certain decisions given observation reports from the affected units as depicted in Figure 1-8. These decisions are a product of conditional probabilities that base such decisions upon the prior existence of certain observations. [17]

The decision possibilities or range of options, expressed as probabilities, for any particular observation must equal one. These probabilities are indicated in the vertical columns of Table 1-1. For each decision, there are a number of observations, expressed as independent probabilities, that may be associated with that decision. Since each observation is an independent event, its probability is not dependent upon any other event. And since there are an indeterminate number of possible events, each of which can take on some finite probability, the total probabilities associated with any particular decision can exceed one. These conditions are expressed in the horizontal columns of Table 1-1.

Therefore, each decision can have a combined probability greater than one, whereas the probabilities for an array of decisions pertaining to any one observation must equal one. Thus, some value assigned to cell $D_2|O_3$ actually represents the probability of decision 2 given the observation set 3 from its respective unit. The probabilities for all decisions (D_1, D_2, and D_3) associated with observation O_3 equals one, whereas decision D_2 is affected by all observations (O_1, O_2, and O_3). This value does not have to equal one.

A further question is whether the observations are valid given that a particular decision is made regarding these observations. The process of calculating the probability of a correct observation given a particular decision is known as posterior, or after the fact, probability analysis because the decision has been made and the issue arises as to whether the observations really supported the decision made. In a network management context, the issue is to what extent the observations are influenced or mitigated by the decision made. This would equate to answering the question of whether a particular sensor is providing worthwhile information needed in decision making. This information can be obtained by evaluating the general expression, $P(O_i|D_i)$, for each cell of the matrix.

The method for finding the probability of an observation's validity given a specific decision is determined from the Bayes' transform:

$$P(O_N|D_i) = P(D_i|O_N)P(O_N) \left/ \sum_{k=1}^{k=x} P(D_i|O_K)P(O_K) \right.$$

where

$P(O_N	D_i)$	= the probability of observation N given decision i
$P(D_i	O_N)$	= the probability of decision i given observation N
$P(O_N)$	= the probability of observation N
$P(D_i	O_K)$	= the probability of decision i given observation K
$P(O_K)$	= the probability of observation K
N	= n th observation
i	= i th decision
K	= K^{th} observation from 1 to x (the total number)

The probability of an event or observation given the decision made is equal to the probability of that decision given the specific observation times the probability of the observation made divided by the summation of the probabilities of all decisions given their respective observations times the probability of their observations.

Table 1-2 shows the same type of probability matrix of decisions versus observations, except that the conditions are reversed. In the case illustrated in Table 1-2, the matrix shows possible observations given the possible decisions that could be reached. As with Table 1-1, for each set of possible outcomes, such as decisions, the total probability for all these outcomes is 1.0 given each possible condition, such as an observation. In Table 1-2, the possible outcomes are reversed with the conditions. Table 1-2 is an example of *a posteriori* analysis where one is interested in the probability of stimuli (conditions) that may have caused a particular conclusion (decision) to be reached.

Table 1-1
Probability Matrix for Possible Decisions
Given Specific Observations

	O_1	O_2	O_3	O_4
D_1	$P(D_1|O_1)$	$P(D_1|O_2)$	$P(D_1|O_3)$	$P(D_1|O_4)$
D_2	$P(D_2|O_1)$	$P(D_2|O_2)$	$P(D_2|O_3)$	$P(D_2|O_4)$
D_3	$P(D_3|O_1)$	$P(D_3|O_2)$	$P(D_3|O_3)$	$P(D_3|O_4)$

Table 1-2
Probability Matrix for Possible Observations
Given Specific Decisions

	D_1	D_2	D_3
O_1	$P(O_1|D_1)$	$P(O_1|D_2)$	$P(O_1|D_3)$
O_2	$P(O_2|D_1)$	$P(O_2|D_2)$	$P(O_2|D_3)$
O_3	$P(O_3|D_1)$	$P(O_3|D_2)$	$P(O_3|D_3)$
O_4	$P(O_4|D_1)$	$P(O_4|D_2)$	$P(O_4|D_3)$

1.5 Decision Support and Executive Information Systems

1.5.1   The Issues for Decision Support Systems

Any hardware/software system that provides information about its operation to the user of that system is a *decision support system*. The term has grown to be somewhat more selective through restriction of its meaning to systems that provide some intelligence to, or analysis of, the raw data collected, so that informed decisions can be made about and actions taken concerning that system. The collection, organization, and even decision-oriented analysis of information pertinent to network management operations is of little value unless human concurrence and control can be exercised over that information. Decision support systems provide this human-oriented capability using highly interactive design tools. Decision support systems were first developed and implemented in numerous ways and places within the military environment over the past 50 years. More recently, the concepts involved have been migrated to the commercial world for use in finance, manufacturing, service provisioning, and distribution arenas, and have begun to be known as executive information systems. [14]

The idea behind the decision support system, as illustrated in Figure 1-9, is to provide the observer with the best view of the situation, resources, actions, and options available to him or her that can be constructed. In this regard, decision support systems are not a substitute

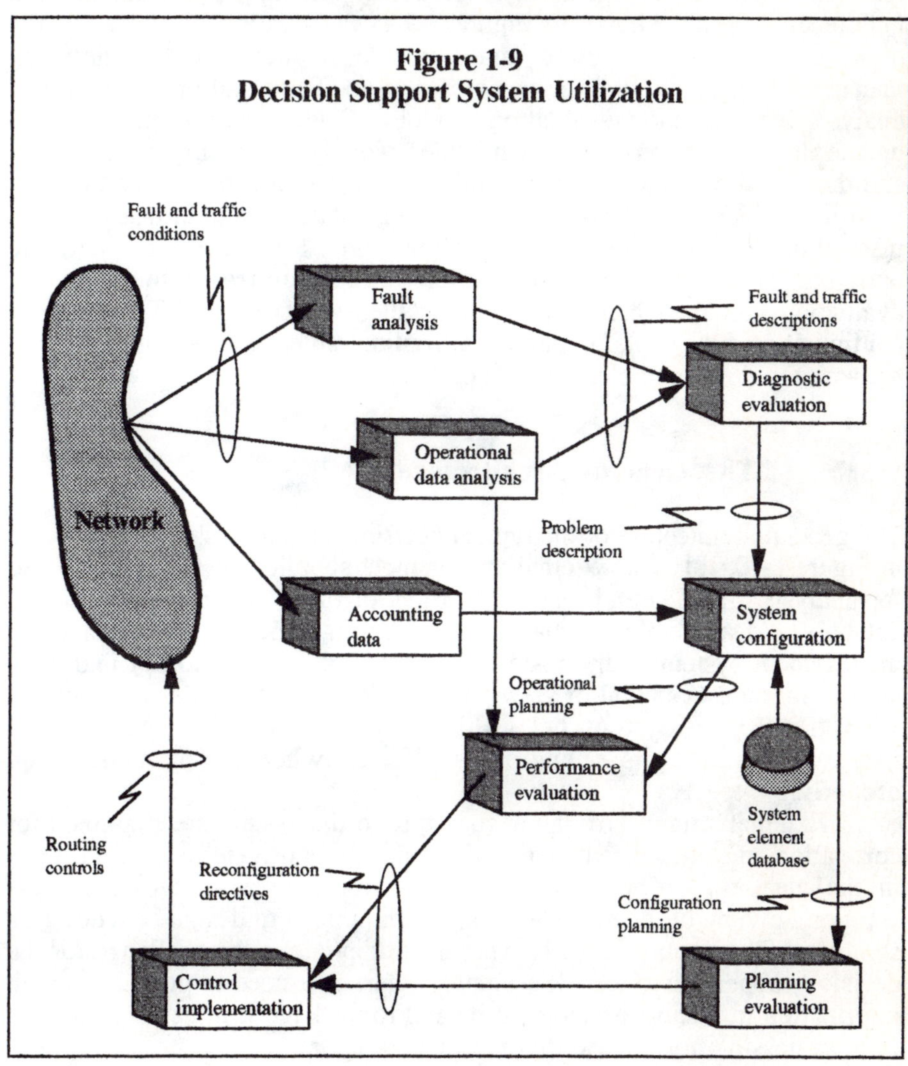

nor alternative for network management facilities, but rather they support the specific five functions defined by the OSI model for the total network management problem. These functions are: (1) fault monitoring; (2) performance monitoring; (3) configuration management; (4) accounting management; and, (5) security management. Each of these will be discussed in more detail in later chapters.

Decision support, however, targets the three main issues within the realm of network management, these being planning, performance, and maintenance. [12] Shown in Figure 1-9 is the relationship between the planning, performance and maintenance aspects of the network management problem and the basic operation. Fault and operational data analysis feed the fault evaluation module. This information is used to update the system configuration module. The system configuration status and the operational data analysis modules feed performance evaluation that in turn provides input to the control implementation module that provides instructions to the system for reconfiguration. Meanwhile, the periodic updating of the system configuration module activates the planning evaluation module which also contains the system goals. The network configuration files contain connection routing tables, equipment inventory, network topology, and element capabilities.

1.5.2   The General Architecture

The general architecture of the typical decision support system is illustrated in Figure 1-10. [8] The external environment supplies the raw data needed for analysis. Trends and historical arrays of these data are stored in the database and available for manipulation through the use of the database management system. The model base provides a replication of reality as seen from the user's point of view. The model is a set of rules (see Section 1.5.3, The Expert System, below) that expresses how the system should function, and how its operation is affected when it doesn't function properly.

External stimuli arriving at the input to the system are examined for correlation with the current status of the model as driven by the data items in the database. The decision support interface assembles the most appropriate view of the situation along with options to design or redesign a solution. To determine exactly what is appropriate is the challenge for the decision support interface. Presumably, there is a need to provide enough information in a short period of time and to package it in as meaningful a form as possible for the user to digest and react to.

It is interesting to note that the technology is and has been available for many years to supply sophisticated decision support systems. Unfortunately, the pressure has not been there to force the development of these tools until recent years, partly due to the lack of competitive influences or lack of complete understanding as to how these tools relate to a company's success. Now that the stimulus is present, there are several problems with which the decision support designer must grapple.

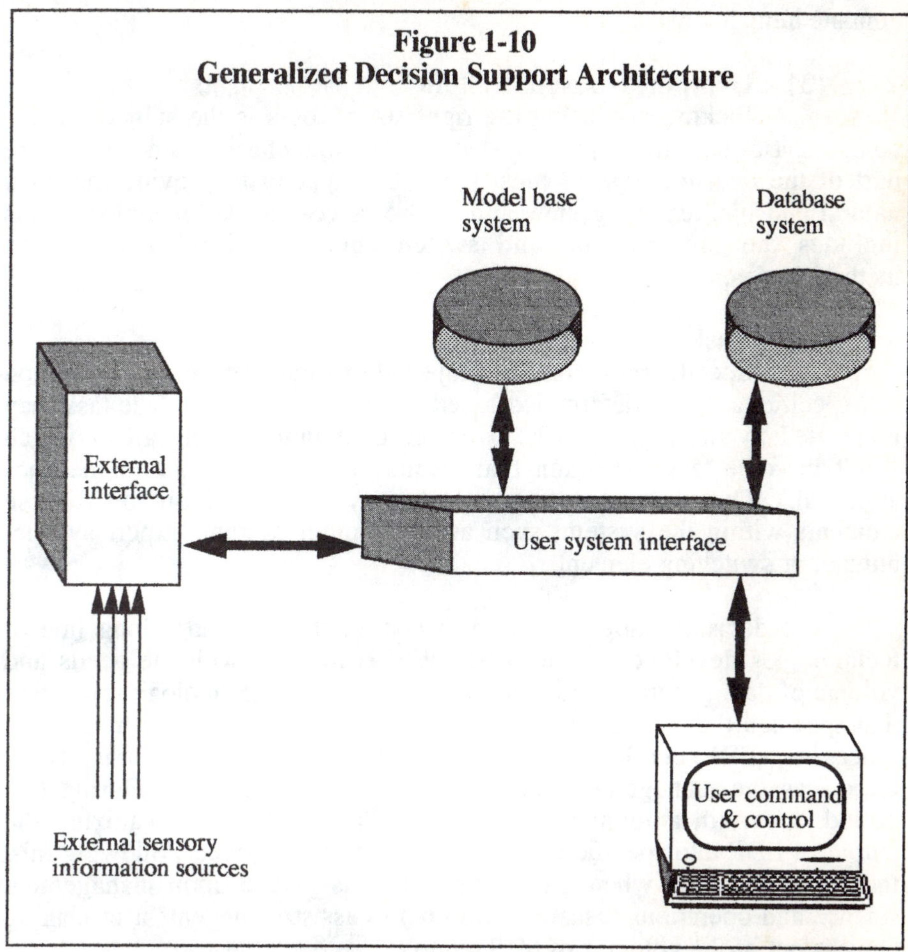

These problems are briefly outlined below:

(1)   Technology

The technology systems needed or desired to solve the network management problem are many and varied. The technology problem usually revolves around the questions of speed, capacity, reliability, physical and functional connectivity, and access to data.

(2)   Behavior

Decision and general psychological theory are areas of investigation that bear directly upon the decision support system and its utility. Decision theory is implemented in several ways, some of which rely on such classical approaches as Bayesian statistics and others that rely on expert systems approaches, such as rule-based methods that attempt to emulate human logic.

(3)   Computer-Based Systems

Picking and using the right set of tools is the subject of this component of decision support systems. The computer instruments that are part of the decision system must have the capacity to provide the most natural and unthreatening view and control access to the operation. This includes the presentation, and switch control and other activation methodologies.

(4)   Tasking Analysis

Specific problems are subjected to an analysis of their situation and specific tasks are determined based upon these analyses. The task may be related to an outage report, overload or planning need, all of which result in some task definition that awaits human review and sometimes approval. The elements of each task may be carried out by diverse elements within the system, such as a computer system, expert system, human, or switching element.

The decision support system is just another step in a long line of technologies developed since World War II to cope with the needs and volume of data generated from and necessary for the technology revolution that spawned the information age. [8] First, there was the electronic data processing (EDP) era that saw the collection of data about processes. Next, there was the management information system (MIS) era. During this period, a certain amount of intelligence was created by organizing the output of EDP into specific reports for management usage. Following this, there was the time when great emphasis was placed upon management science and operations research (MS/OR) to assist management in making choices out of the volumes of MIS reports.

MS/OR helps one make decisions from highly structured information. Finally, there is now the era of decision support systems (DSSs). This is a set of technologies that allows the user to design his or her own solutions based upon incomplete information. This thrust provides a very fertile ground from which to implement many forms of intelligence technologies for the purposes of emulating the human brain into effective DSSs.

1.5.3   The Expert System

Expert systems are a major element of decision support systems because they provide the intelligence needed for the decisions reached. They can be roughly equated to the model(s) used internally to aid the decision process. Expert systems have three main features, as shown in Figure 1-11, these being:

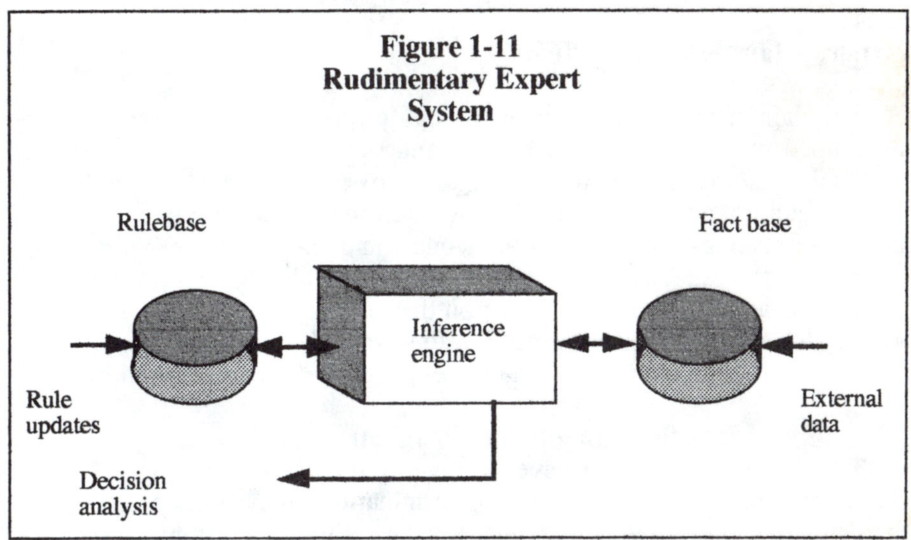

(1) The Inference Engine

The inference engine is that software program that uses the production rules to work out the logic of a particular problem. This is done mainly through sophisticated sequential searching and comparisons using lists supplemented with external facts. Additionally, the conclusions drawn are weighted by probabilities of success or correctness. Thus, facts are collected, hypotheses accessed, probabilistic judgments developed, and conclusions drawn that fit the initial facts.

(2) The Rulebase

The rulebase is that segment of the database that contains the rules upon which the inference engine operates to resolve the problem at hand. The rule base is essentially a model of the system in which resides a representation of the system's structure, status, and history. The rulebase can be invoked to reveal the rulebase's formal logic, and its and parameter assignments.

(3) The Factbase

The factbase is that portion of the database that provides the basic facts or data used by the inference engine to evaluate the rules, and therefore, to reach a conclusion. This information usually comes from external means, such as from attached or remote sensors, or from organized sources reached over communications channels. The structure of the factbase is extremely important to the effective use of the data available.

Expert Maintenance Systems

Expert maintenance systems were probably the first expert systems to be developed with computer technology and the decision support interfaces for such systems were and are, needless to say, important as well. The terms maintenance, fault control, and failure monitoring are sometimes used interchangeably to imply the same thing, i.e., that the maintenance system has the capacity to detect, identify, and resolve system problems.

The decision support application that interfaces with the maintenance system has several requirements to satisfy in order to make it appropriate for the job at hand, and these requirements are summarized as follows:

(1) Identification of the Problem

The DSS must have access to the data necessary to determine that a problem has occurred. This determination many times surfaces as a result of several indicators as opposed to a single occurrence. Problem identification may come in many forms, such as failure of a parametric test, or a logic test.

(2) Isolation of the Problem

Determining where the problem exists is many times not an easy task. The location may be masked by several other conditions or situations that make this step a difficult one. Testing for a specific location is sometimes an effective tool.

(3) Association of the Problem

The exposure of the problem immediately stimulates a search of the previous times or situations during which the problem may have occurred. If no such circumstances have occurred, association can be established by reference to an operational model of the system. Such events and references provide a starting point for problem attack.

(4) Implications of the Problem

The association of the problem with its place in the system often provides information that might implicate the problem with other other system elements that the current problem may infect. The current problem may also imply or implicate other processes that are more pertinent to the malfunction or bottleneck situation.

(5) Visualization of the Problem

A conceptualization of the problem as something that a human can appreciate is important in any system, expert or otherwise. Such a concept presentation can be provided as a graphical display or even a functioning picture of the system. The problem visualization is executed by simulation of the system functionality or processes and as a result will often suggest one or more solutions to the problem at hand.

(6) Options for Solving the Problem

The system simulation, isolation and association are the principal vehicles for enumerating the options for problem resolution. These processes provide a list of possible options that can be implemented. In addition, the options may carry a list of caveats and degrees of difficulty that may be encountered as a result of exercising each of these options.

Performance Management Systems

The performance management system likewise has several uses for decision support. Such systems monitor operational traffic and, using decision support capabilities can anticipate bottlenecks, isolate operational trouble areas, and correlate fault messages with operational conditions. Examples of performance management systems implemented include NEMESYS and XTRAC. [5] NEMESYS is an AT&T tool developed to assist in fighting congestion in the inter-exchange carrier business. It is based upon techniques to optimize performance during overload or outage emergencies. Information on the number of and reasons for call blockages is collected and used to derive suggestions for call rerouting. XTRAC is a system that also reallocates resources based upon priority considerations.

Performance measurement and problem resolution based upon performance problems, are both associated with the types of performance problems incurred within the network whether they be location specific or time based. These types of problems can be categorized as follows:

(1) Routine Capacity Overload

This is the most common problem handled by a performance management system. It occurs as a result of some cable or transmission route becoming undersized for the load placed upon it as traffic increases over that route with time. In this type of situation, the route experiences periodic durations of overload at an increasing frequency and for an increasing duration per overload period as the general loading becomes greater.

(2)   Focused Overload

The focused overload results from sudden increases in local demand due to some type of emergency either manmade or natural. Examples might include severe weather or a nuclear power plant problem. The focused overload becomes most troublesome when the the facility needed to complete the circuit itself is the object of the overload, such as the telephone end office, because in this case, there is little or no rerouting possible around the problem.

(3)   General Overload

In this type of overload situation, the overload is not localized as with the focused overload. It is ubiquitous throughout the network. An example of a general overload situation would be Mother's Day or Christmas. During a general overload period, there is little rerouting and few alternatives available to resolve the problem. The problem will usually go away in a few hours. Recent catastrophes, however, have shown that general overload may not go away in a few hours. Certain local problems such as that in New York City that affected the entire air traffic system and the software bugs that affected new telephone switches around the country are problems that may be harbingers of more dire situations in the future.

(4)   Facility Failure

Facility failure is a situation where there may be a failure of a trunk group as a result of weather problems. Here the situation resembles a focused failure except that the capacity has been reduced due to equipment failure as opposed to an increase in demand. In many such cases, the problem can be circumvented by use of spare components or switching around the problem.

1.5.4   Planning Management Systems

The typical planning function occurs either to support cost, schedule, or technical performance objectives. One example of a planning management system is one which operates as an expert assistant in designing networks. It is called *DESIGNET* and acts as a configuration tool for boards, racks, cables, interfaces, etc., at each site in a network environment. [5] The DESIGNET module provides the user with many different views of the network configuration and performs sensitivity analyses by answering why and what-if questions for the operator. Another such planning system was developed at GTE Laboratories and is known as *MARVEL*. [9]

1.5.5   Examples of Executive Information Systems

As the technology becomes ever more capable of supporting sophisticated information needs, the possibilities for commercial decision support systems, referred to as executive information systems, approach ever more rapidly.   The Frito-Lay company is one such example of effective technology transfer from military battlefields to commercial battlefields[4]. See Figure 1-12 for a portrayal of this system.

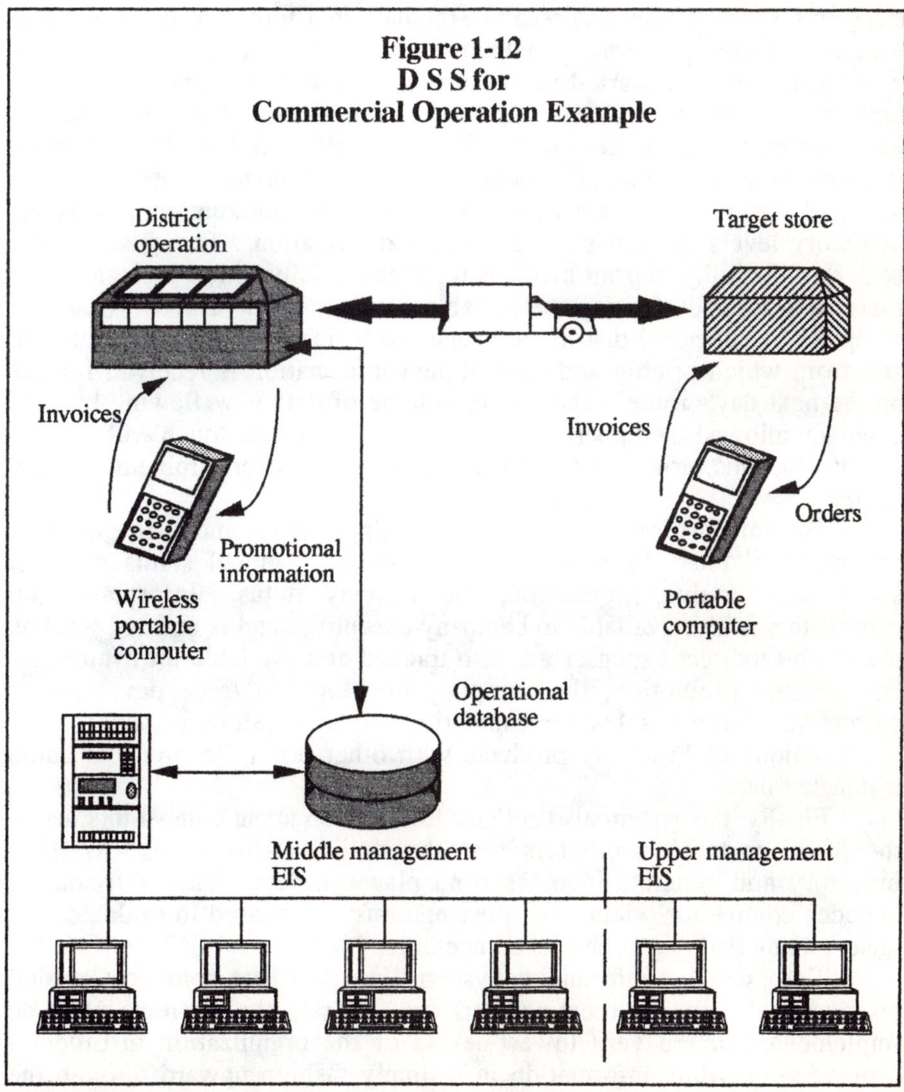

Frito-Lay has developed a distributed, layered executive information system that provides this snack company food for action as well as thought. The system is available in at least two flavors for both the upper and middle management teams so that they can track, plan, and execute their operations more effectively. The main objective of this executive information system is to carve up the massive array of constant data flowing to the operation into small enough pieces that can be digested by even the most modest of the desktop systems at the levels indicated. Each targeted display station also receives the data in a form that is concise and focused upon the concerns of that particular station user.

The company started at the bottom of the management strata and built the information system upward. The system has two types of information, that pertaining to the Frito-Lay products and that pertaining to competitors' products. The basic information gathered is obtained from the field supply trucks that collect and enter into portable computers the inventory levels at each store and order information. The systems also have the capability to print invoices for product delivered at each stop on a route. At the end of a day the portable computer is interfaced to a larger computer at a regional distribution center to which the invoices are copied, and from which pricing and promotional information is received for use on the next day's route. The timely volume of data now flowing into the company allowed the upper management to be in closer touch with current operations, thus providing them with the means to avert problems as they started.

The information now available to upper management, in graphical and pictorial chart form, included reports on regional status, product status, sales status, production, and delivery status. Information on competitors is also available to company executives and is updated weekly. Direct and indirect expenses are also tracked and available for evaluation. For product promotion, the company developed a trade development system that allows sales representatives to show store managers how combinations of Frito-Lay products with other non-Frito products could stimulate sales.

Finally, the system also collects information about competitors, using the same portable computers that the delivery drivers use to track inventory and generate invoices from placed orders. Such information includes competitor pricing and presentations and is used to evaluate how the company is doing versus its competitors.

The executive information system illustrated here points out several lessons that can be recapped briefly. First, the system must be implemented at the very lowest levels of the organization in order to provide worthwhile information in a timely fashion upward through the organizational structure. Second, the information provided must be presented in a way so that it can be made useful to the various management tiers requiring it, such as charts, graphs, and characterizations. Third, the information must be segmented to the maximum extent practical so that very specific answers can be extracted for very specific questions. And, fourth, the data must be organized and presented differently for the different levels of management using it so that proper understanding can be maintained.

1.6 A Look Ahead

This first chapter was a summary of the main topics to be discussed further in this book. These main topics include sensors, data fusion, and decision support or executive information systems. Each topic is intimately associated with one or more leading edge technologies one of which is expert systems. These systems form the common threads in this book as a whole. Another, but more common, technology is one which we will describe as systems integration. Here *systems integration* is the tying together of elements in a way that is meaningful. Accomplishing this task requires the implementation of different analytical techniques, hence the characterization of system integration as a technology.

In this book we will survey and explore the various elements of the network management system that can be implemented with the technologies needed to implement any of these smart features. To begin with, there are several areas in the network system into which smart systems technology can be infused. The first area is the front end monitoring of system components. Sensors monitor, evaluate, and report on component performance and status. Since the system sensors, as we will characterize them, are operating with the network components in locations remote from the monitoring and control points, we would like for them to be somewhat autonomous in their actions. The various functioning components and elements of the telecommunications network are usually the sources of status information about the network that is needed for the assessment of the total system. Each is a sensor of sorts that collects and reports on the status of its environment.

The second area into which smart systems technology can be infused into the network is that point at which the sensor data are typed, sorted, and categorized. We will refer to this as the *data fusion point*, and the overall process is referred to as simply *data fusion*. Information flowing into this area must be continually evaluated both individually and in relation to other data in order to track the status and performance of the system. Data fusion is not as simple as it may appear. The data are many times from vastly different environments and from different vendor equipment.   The formats vary from each other and the correct interpretation of data contents can be very different. Data fusion is also the point in the data collection and analysis functions at which the confidence of incoming status data is confirmed or suspected. As a matter of fact, the major reason for implementing data fusion is to verify status indicators.

The message behind any techniques and/or technologies used to further the aims of network management is what directly impacts the user. At its simplest and most personal level, this is the paradigm of an operator sitting at a terminal and pushing the return key after typing in a request. Nothing happens, and the user doesn't know what to do, or what happened, or where to start to resolve the problem, if there is one. Smart systems should be developed and implemented so as to answer these concerns. These systems should be able to isolate the problem, take corrective action, restore the user's session activity, notify the user what happened, and advise the user as to any actions that the user may find helpful at that point. If smart systems can assist in these objectives, they will have served their purpose.

Summary:

Smart systems are really intelligent components of the overall network management system. Smart systems can be divided into three major sets: sensory, fusion, and decision support. Each of these, in turn, can be realized through the implementation of leading edge technologies, such as neural nets, expert systems, and robotic features. The system sensors collect the data; the fusion process integrates the data collected; and, the decision support process makes sense of the fused data. The proper implementation of any of these features into a network can have tremendous positive impacts upon the simplicity, quality, and operating costs of the target system.

In the near future, smart systems will become much more than just advisors. They will begin to be integrated into the decision and control fabric of the typical network management system as such systems become more and more complex. Already telecommunications manufacturers are beginning to embed expert systems into their products, and smart system shells are being developed for the telecommunications applications of monitoring and diagnosis. Such applications will include alarm monitoring, data filtering, automatic rerouting, reconfiguration, and network performance monitoring. Finally, certain technology and quality issues will begin to be addressed in the near term as smart systems technology is directed at video teleconferencing, high-definition television (HDTV), where data noise filtering and compression are applied to these new technology needs.

References

(1)   Bertuol, B., "Sensors as Components for Automotive Systems," *Sensors and Actuators A*, 25-27 (1991), pp 95-102.

(2)   "Digital Input/Output Local Area Network Node Card," *IBM Technical Disclosure Bulletin,* vol. 28, no. 10, March, 1986.

(3)   Ericson, C., L. T. Ericson and D. Minoli (eds.), *Expert Systems Applications in Integrated Network Management*, Artech House, Inc., Norwood, MA, 1989.

(4)   Feder, B. J., "Frito-Lay's Speedy Data Network," *New York Times,* November 8, 1990.

(5)   Goyal, S. K., and R. W. Worrest, "Expert System Applications to Network Management," in *Expert System Applications to Telecommunications*, Liebowitz, J.(ed.), John Wiley, New York, 1988, p1.

(6)   Harris, C. J., and I. White (eds.), *Advances in Command, Control & Communication Systems*, Peter Peregrinus Ltd., London, UK, 1987.

(7)   Henning, W., "Bus Systems," *Sensors and Actuators A*, 25-27 (1991) pp 109-113.

(8)   Hoffman, M., "Technology Profile Neural Networks," *Techmonitoring*, SRI International, July 1991.

(9)   Hopple, G.W., *The State of the Art in Decision Support Systems*, QED Information Sciences, Inc., Wellesley, MA, 1988.

(10)   Lemmon, A., *Marvel - A Knowledge-Based Planning System*, GTE Laboratories, Internal Report, 1986.

(11)   Loomis, M. E. S., *Data Management and File Structures*, Prentice Hall, Englewood Cliffs, NJ, 1989.

(12)   Singleton, W. T., "Man-Machine Aspects of Command and Control," in *Advances in Command, Control and Communication Systems*, Harris, C. J., and I. White (eds.), Peter Peregrinus Ltd., London, UK, 1987.

(13)   Wagner, U., "Reliability and Fault Tolerance of Low-cost Multipoint Sensor Interfaces," *Sensors and Actuators A*, 25-27 (1991) pp 73-78.

(14)   Walker, T. C., and R. K. Miller, *Expert Systems 1990: An Assessment of Technology and Applications*, SEAI Technical Publications, Madison, GA 30650.

(15)   Wallach, R. M., "The E.I.S. State," *Computer Systems News*, March 3, 1990, p 49.

(16)   White, F. E., and J. Llinas, "Data Fusion: the process of C^3I. (command, control, communications and intelligence)," *Defense Electronics*, vol. 22, p 77, June 1990.

(17)   Wilson, G. B., "Some Aspects of Data Fusion," in *Advances in Command, Control and Communication Systems*, Harris, C. J., and I. White (eds.), Peter Peregrinus Ltd., London, UK, 1987.

Chapter

2

The Functional Requirements of Smart Systems

Chapter Highlights:

Smart systems technology as it relates to network management embraces the precepts and standards associated with the OSI network management model. In this chapter we will review the five major functions associated with the OSI model and discuss some of the opportunities available to smart systems as they might apply to these five functions. The underlying assumption in this chapter is that smart sensors, decision support systems, and data fusion are all involved in the solutions and opportunities for smart systems to be proposed. However, the specifics of how these technologies are to be applied will be left to succeeding chapters. The level of definition concerning these five functions is fairly well established at this point, thus providing more definition to the opportunities available.

2.1 Orientation

This chapter describes the five network management functions as they relate to and influence the smart systems developments within these areas. Without these categorizations, the consistent design and implementation of smart systems within the network management environment would be difficult, primarily because they would have no boundaries or defined responsibilities within which these smart systems would operate.

The functions of network management were defined by CCITT in the mid-1980s. [10,12] Since that time, each of these five network management functions has been individually defined so that a consistent set of tasks associated with each has emerged in the standards arena, and these tasks are briefly reviewed here. This chapter will also discuss the implications of these functions in the light of intelligent features. [14] For instance, the fault management function is especially adaptable for implementation within the context of expert systems algorithms.

Any time a new product, service, or system is contemplated, the customer needs, or assumed needs, must be determined in some reasonable manner. There are many sophisticated methods of getting at such needs in varying degrees of granularity, all of which are used to get at the required system configuration. The U.S. military has spent many decades and billions of dollars refining these processes, which today are generally referred to as the *requirements definition* and *system specification* processes. During the requirements definition process, it soon becomes quite clear that the various functions that make up this system can be accomplished through varying degrees and configurations of automation. This same approach is applicable to network management, where user needs are evaluated against the available technologies to yield the target system, that also has varying degrees of possible automation.

At present, network management development efforts seem to be predominantly concerned with performance and fault management. However, emerging technologies and increased uniformity of product features will cause this preoccupation with performance and fault management to balance out with increased attention to such issues as remote commanding of the network elements, along with network security and system accounting. These features will become increasingly integrated into the typical network management system's list of capabilities.

First, we will discuss the reason for addressing the five network management functions. Next, we will explore the five Comite Consultatif

International Telephonique et Telegraphique (CCITT)-approved functions associated with network management, which are [13]:

    Fault management
    Security management
    Performance management
    Configuration management
    Accounting management

After this, we will consider how these functions can be accomplished through varying degrees of automation, and particularly through expert and neural systems implementation. In summary, it is clear that the variability of these functions affords many opportunities for sophisticated automation which potentially represent significant cost savings, high quality product and service offerings, and viability for offerings in the marketplace.

2.2 Operational Problems and Solutions

Before any discussion of requirements for network management can be discussed or resolved, we must identify the problem or problems. Action of any sort, and in any situation, is inappropriate unless we can identify a need or, said another way, until a problem is isolated. In the case of network management, the reason why the five functions of network management were defined was because a series of problems or potential problems were identified for monitoring and subsequent resolution. Such problems include the need for fault detection and isolation, inventory status, security system operation and reports, recordkeeping for the usage requests, both satisfied or not, and current indicators of system usage. For instance, the excessive delay in reaction times between a request to the network and an answer may indicate one of several types of problems, such as an operational congestion problem, an equipment outage problem, or a software problem.

Figure 2-1 illustrates this situation. A delay realized at either end of the network links could alert the system administrator that a problem has occurred somewhere in the system. [1] The smart systems discussed later could be the instruments of isolating and treating these problems. As an example, a delay could be indicative of a congestion problem, an outage problem, or switching delay problem. The smart systems may be exposed to all of these for resolution.

Once a particular operational problem has been identified, it must be addressed so as to alleviate the effects of this problem. A good example of the type of problem or set of problems that can be addressed readily is that associated with the physical layer management of the system. [3] The solution for a problem within the realm of network management starts with the description of the proposed solution, sometimes referred to as an operational concept document. Such a document spells out the solution in terms of the problem, alternative approaches to the solution, impact of the solution, cost, and cost/revenue benefits if any. The typical format for such a document is as follows:

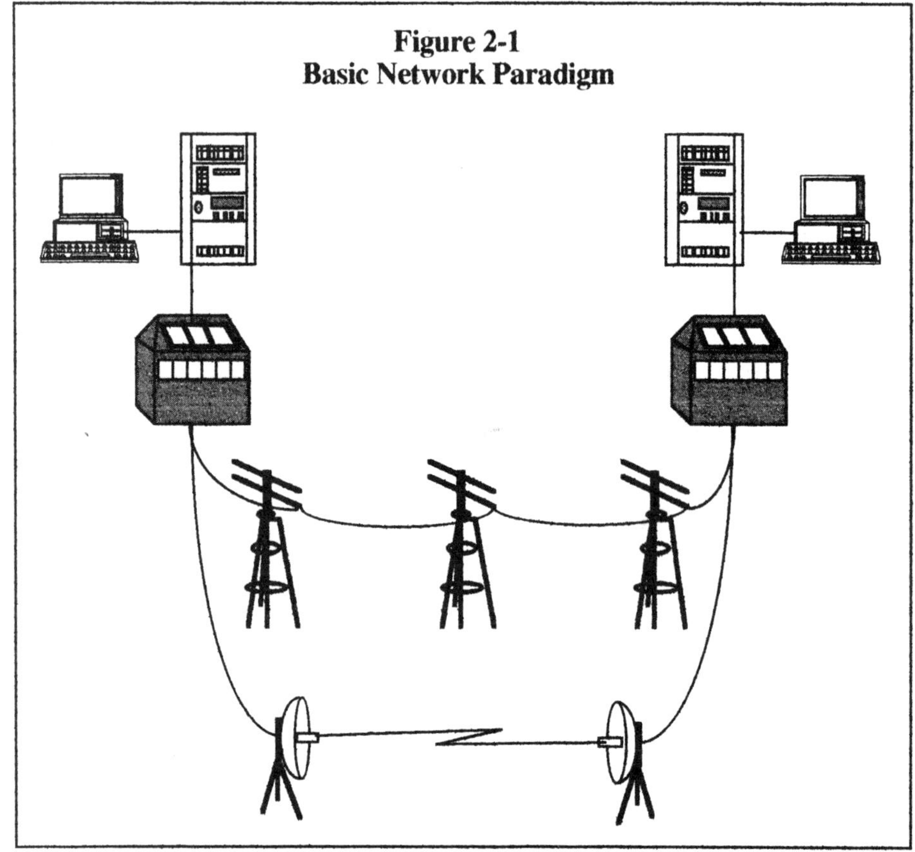
Figure 2-1
Basic Network Paradigm

(1)   Statement of the problem
(2)   Identification of alternative solutions
(3)   Description of the selected solution and why

(4)   Operational impacts of the solution, e.g., cost, revenue, operational disruptions, technical issues, schedule impacts for other planned changes, and management of customer perceptions

Once this set of steps has been followed, the relative wisdom of going through the process to achieve the results called for in the operational concept can be better evaluated and prioritized for action planning.

2.3 The Network Management Functions

2.3.1   Fault Management

Definition

Fault management can be defined as the real-time or near-real-time monitoring of the elements of a computer communications network with attendant resolution of related problems. This function addresses both hardware and software activities that are part of the system operation. These elements are both physical and functional, and can include connectivity errors, equipment failures, performance bottlenecks, and performance discontinuities. [12] Fault management may obtain some of its inputs from other functions, such as configuration management, but the evaluation of inputs is irrelevant to their source.

Description

The fault management function interfaces with the various system components sometimes directly, as with network elements, or sometimes indirectly via subnetwork managers, such as LAN managers. [5, 7] In the OSI scheme for network management, the fault management function, as with the other interfaces with the network through the Systems Management Applications Entity (SMAE), that is located at each node, via the Common Management Interface Protocol (CMIP) gateway. A node may be anything from a single piece of routing equipment to a complex array of data processors. The key feature is that the node must somehow terminate the incoming data channels and reconnect them at the outgoing side.

The fault management function detects faults by receiving signals that are concerned with certain functional and physical errors relating to the system operation (detection). [4] Detection is an age-old problem upon which billions of dollars have been spent over the years, and to which decades of time have been devoted. Detection is rarely equated to absolute values, but rather is declared on the basis of comparisons, thresholds, or other forms of relative value. More often than not, detection of the object signal is expected in the midst of noise, not to mention other conflicting and confusing signals, in addition to the one being anticipated or sought. The signal-to-noise ratio provides the basic parameter for analysis, and when coupled with the detection threshold and the actual measurement collected, provides an indication of the relative probability of the existence of the signal of interest. In a digital environment, the report of a detection may be a formatted message, but its origin can usually be equated to the analog parameters or large digital samples that create the detection problem cited above. Figure 2-2 illustrates the basic issues involved in signal detection.

The cross-hatched area labeled as the probability of detection, P(D), represents the probability of detection of a fault, given that a fault has actually occurred. Sometimes, a fault is declared when, in fact, such is not the case, and this is represented as the probability of a false alarm, P(FA). Both of these probabilities are predicated upon the threshold of detection established for that specific situation. As the detection threshold is moved back and forth from left to right and back again, the probability of detection is improved or degraded, but so is the probability of suffering from more or fewer false alarms. The curve labeled signal includes all correct indications of faults as they occur.

The curve labeled as noise includes all non-fault or false alarm reports. In some cases, signals that pass the criteria for declaration as actual fault events are, in fact, not faults at all. Such events are false alarms. Depending upon where the detection threshold is placed, the likelihood of experiencing more or fewer such false alarms is altered. As the detection threshold is shifted to the left, there is a higher probability of experiencing false alarms. The positive aspect of moving the detection threshold to the left is that there is also a higher probability of detecting all actual faults.

The trick is to separate the noise and event curves as much as possible in order to maximize the detection capability while, at the same time, minimizing the possibility of risking false alarms. In practice, this can be done by improving the quality and intelligence of the sensors involved, performing certain filtering and analysis functions in the data fusion function, and providing the intelligence in the decision support module(s) to discriminate between actual and false reports.

Following the detection of faults, the fault management function makes an analysis of these signals and determines the offending component (diagnosis) down to some level of ambiguity or indifference. The term *indifference* is used here to indicate that certain levels of fault isolation may be beyond the capacities of the current diagnostic system.

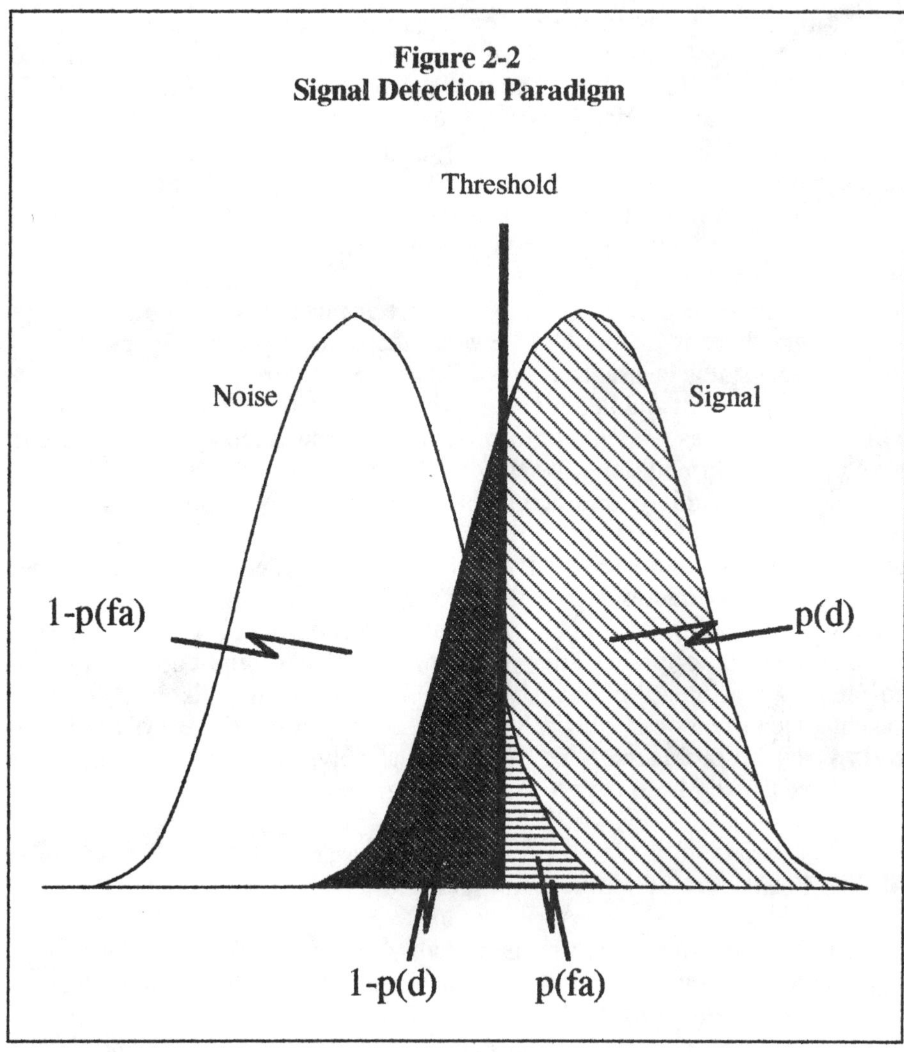

Many times, status signals corresponding to errors can only be traced to two or more physical or functional components or both. Such situations give way to varying degrees of ambiguity. If the error signal is coarse, or highly diffuse, in nature (i.e., it represents the operation of several components) ambiguity arises because of the lack of more detailed information. If the error signal arises because the operation of the components is intricate and complicated with feedback or multipathing of signals, ambiguity arises because of simultaneous possibilities of operational failure(s). Figure 2-3 shows the conceptual causes of ambiguity.

In some systems, status information may come over separate channels, or may be interleaved with the operational data. In Case A, the accumulation of information about the status of the system elements is spread over n number of elements, thus confusion may arise as to which elements are at fault when all status reports are processed. Even though status reports are usually identified with the equipment source reporting, failures in one piece of equipment may ripple through the data stream to affect equipment in other parts of the system. In Case B, possible simultaneous failures, or single failures that affect all channels at the same time, may confuse the deciphering of status messages. Ambiguity may arise where there is decoupling between the data traffic configuration and the incoming status messages.

Finally, fault management provides assistance in fault correction. This support takes two major forms, namely the reinstatement of some previously performed state, or the tracing of events that is facilitated by event logging. Adding, and being able to access analysis points, as needed is an old method of tracking and isolating trouble points. This is a form of reinstatement as indicated above that provides added identification and therefore fault correction. Event tracking also aids in the fault correction process by isolating the points of failure, those points at which corrective action must take place and, therefore, provides indications of the corrections that must be accomplished. It appears quite likely that smart systems technology can play a large part in the correction of system faults and associated problems. The tools used to solve such problems, up until now, have usually been expert systems programs.

Major Functions of Fault Management

The fault management function is probably one of the first functions to be implemented in any network management environment. The objectives of smart systems implementation are to provide effective and cost-efficient solutions to the data collection problem as well as the accurate and timely interpretation of the data collected. These objectives, hopefully, can be implemented using sophisticated technology to minimize the costs to the point of being optimized for the result desired. Ambiguity in the messages flowing across the CMIP gateway makes the job of accurate and timely interpretation difficult at best without the proper tools.

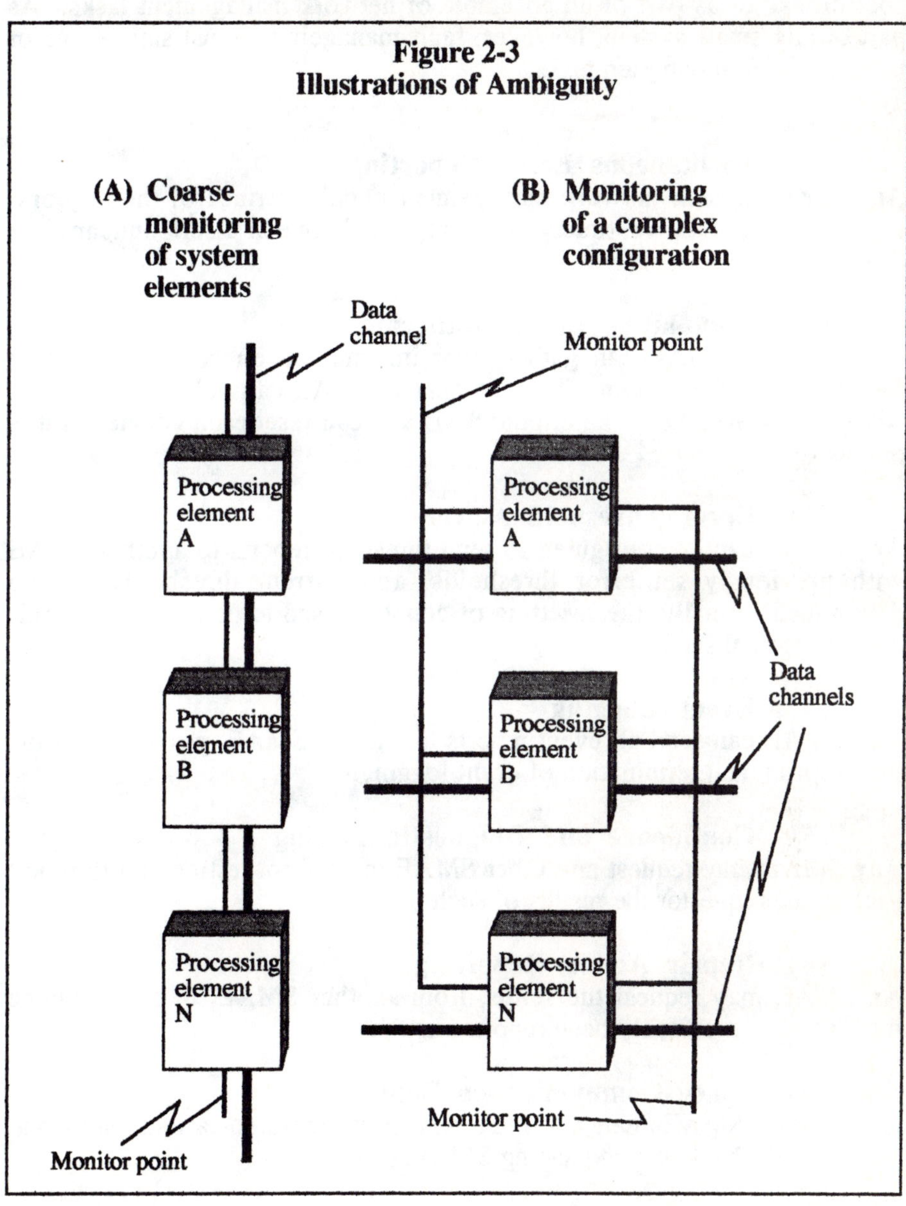

Such tools are selected based upon the specifications of the system to be supported. As one of the primary functions of network management, fault management may be implemented as either a standalone network control task or as part of an ensemble of network management tasks. As part of any smart system, however, fault management must satisfy one or more of the following ten tasks:

(1)   Spontaneous Error Reporting

An SMAE, at each network management nodal interface of the network, can send and receive timely error reports between itself and another SMAE.

(2)   Cumulative Error Gathering

A designated SMAE can gather error information on behalf of another SMAE within the system. The designated SMAE can poll error counters within other SMAEs on a periodic basis and can reset each counter as it is polled.

(3)   Error Threshold Alarm

Any SMAE can be configured to send threshold reports to another SMAE with previously set error thresholds, and current thresholds can be determined. Finally, the resetting of counters used to compare thresholds can be accomplished.

(4)   Event Logging

Any SMAE can send all event reports to another SMAE, providing for the initialization and termination of event logging.

(5)   Confidence and Diagnostic Testing

Any SMAE may request any other SMAE to perform testing and to report back to the requestor the results of such testing.

(6)   Repair Action Reporting

An SMAE may request the status, from another SMAE, of any resource that has been previously been reported as faulty

(7)   Trace Communication Path

Cooperating SMAEs can test interconnecting communications paths and report results back to a requesting SMAE.

(8)   Resource Reinitialization

An SMAE can request another SMAE to set the initial state(s) of some resource to a known parameter(s).

(9) Event Tracing
One SMAE can request another SMAE to start or stop logging specific events locally, and to report back the status of this exercise.

(10) Fault Management Information Gathering
This facility provides for one SMAE to collect, dump, and analyze local information so as to support other SMAEs making such requests.

The tools referred to above that are needed to support these tasks could include the use of triggering mechanisms within the SMAE software to initiate any of the processes cited above, and possibly, based upon the development of smart systems technology, to facilitate the completion of these tasks. Additionally, such tools could include automatic data collection in anticipation of such a request from another SMAE that would be based upon traffic and/or other operational considerations.

Faults and Failure Mechanisms

The nature of anticipated faults, as the absence of functional capabilities, influences the design of equipment and the functions pertaining thereto. [2] Failure, on the other hand, can be expressed as the lack of expected performance. Faults are characterized differently for digital circuits as opposed to analog circuits. For digital circuits, faults are categorized according to their effects upon the logic values of the circuitry, rather than according to their direct causes upon the functionality of the system. The following are types of faults that may be encountered from time to time.

(1) Single Stuck-at Faults
These faults occur due to an inability to make a transition between logic levels in some circuit. This is usually caused by alternate failures within the circuitry or some type of deposition of material during the manufacturing process.

(2) Multiple Stuck-at Faults
These types of faults may be the result of manufacturing defects caused by over-etching, an incorrect PC layer, etc. For non-interactive faults in this category, tests can be devised to expose each separately, while interactive faults may not be revealed with standard tests.

(3)   Bridging Faults

These faults can be thought of as variations of the single and multiple
stuck-at faults. Foreign material touching two or more component parts or
circuit traces is the most common form of bridging fault. A not-so-elegant
approach to such problems has been to implement trace cuts as a fix.

(4)   Intermittent Faults

Any fault that occurs and then disappears without intervening action taken
to correct the fault is an intermittent fault. Such faults are usually
discovered by examining the physical implementation of the unit in
question. Vibration is a likely cause of intermittent faults, as is thermal
stress.

(5)   Memory Faults

Chip miniaturization with its constraints upon separations is a major cause
of such faults. Tighter and tighter constraints brought about by transitions
from LSI to VLSI to VHSIC technology where the separations between
traces is 0.5 microns or less, causes such faults to be more common.

(6)   Time-Dependent Faults

Such faults are usually linked to physical rather than electronic roots.
Varying the memory refresh times of a display monitor are an example of
such faulty conditions that can be attributed to physical conditions.

Implications for Smart Systems

Expert systems technology is directly applicable to the fault management
function. Any intelligent systems technology has two basic features,
namely to predict and to identify. Expert systems can aid in the prediction
of potential problems, and additionally, in the identification or isolation of
problem sources. Such expert systems implementations can be implanted in
decision support systems stations manned by the system operators.

2.3.2   Security Management

Definition

Security involves the unequivocal knowledge of the persons, places or
things involved in the conveyance of information from one locale to
another and the protection of that data from unauthorized receivers or
senders, or the corruption of the data. Communications security (comsec)

management requires that the necessary entities involved in the information storage and transfer process:

> (1) be aware of information that may be of use to the intended recipient

> (2) have access to the necessary information storage facilities

> (3) be aware of the attempted transfer

> (4) be capable of the transfer

> (5) be authorized to participate in the transfer

> (6) be assured that the receiving parties have a need to receive the transfer

> (7) be assured that the transfer is that intended by the originator(s)

Security management must be involved in all of these aspects of the information storage and transfer process if proper security is to be observed. There are five main security services necessary for effective security management, these being authentication, access control, confidentiality, integrity, and nonrepudiation. The allocation of security services to the basic system elements of communication, i.e., the sender, receiver, and system features, is shown in Figure 2-4.

(1)   Authentication
Authentication is the verification of the fact that the information conveyed is actually coming from the source or authority expected. As implied, the authentication can come from both peer level or from some higher authority.

(2)   Access Control
Access control is the ability to restrict access or access control to some source. Access control may be exercised for facilities, services, storage repositories, data structures, or communications channels.

(3)   Confidentiality
Confidentiality is a property of data that it cannot be made available to any sources or processes without specific procedures in effect.   These procedures specify the receiver's need to receive the data conveyed,

authority to release the data, protection of the data from disclosure, verification of receipt, and authentication of the sender.

(4)   Integrity

Integrity is the verification that the data received is, in fact, received as transmitted.   Either intentional or unintentional alterations in the data content could have far-reaching consequences for both the transmitter and receiver.   This service provides security to the data after it has already been collected, and passed the first three filters.

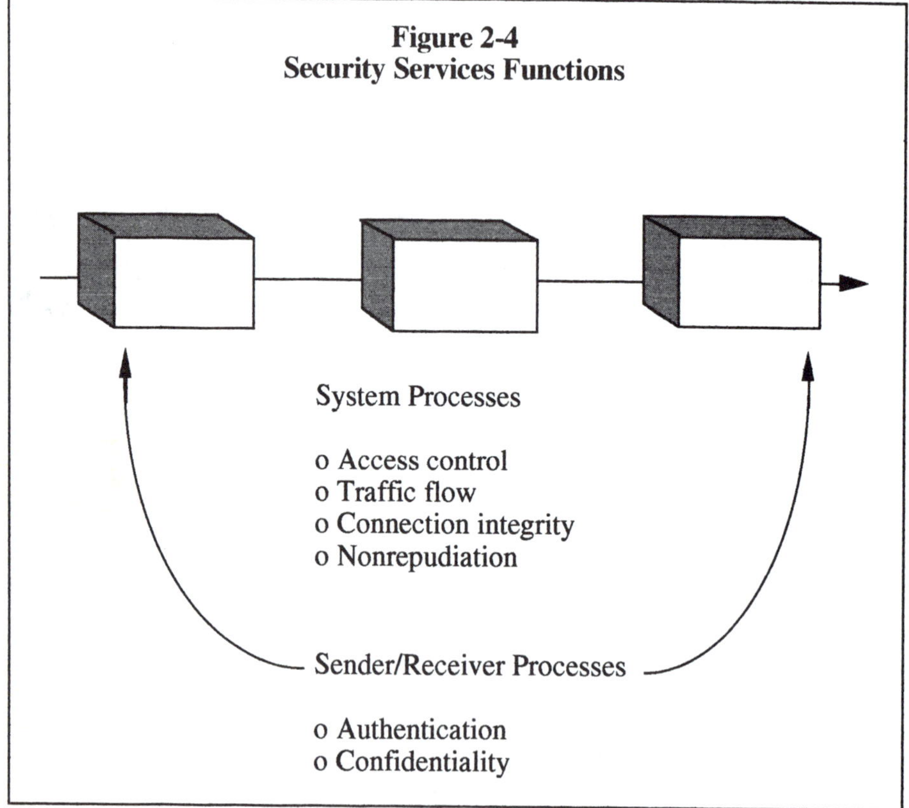

Figure 2-4
Security Services Functions

(5)   Nonrepudiation

Non-repudiation is the verification that either the originator or receiver or both did actually send or receive the transmission in question, so that neither party can later deny the transmission or receipt of the data.   This service is often overlooked by security systems.

Description

Security management can be further delineated as roughly 10 different sub-services necessary to assure maximum system security, these being:

(1)   Data Origin Authentication

Data origin authentication is the corroboration of a judgment that the data received is the same as that intended in the transmission. Many times the receiver may not be absolutely sure that the information received is that intended by the sender. In such cases, an additional indication of the sender's intentions is sent to the intended receiver to reassure the receiver that, in fact, the sender did intend the data signal.

(2)   Peer Entity Authentication

Peer entity authentication is the corroboration that the peer or equal in a configuration is the one intended. This service involves the assurance that the various elements in the communications transmission scheme are the ones claimed. If the communication elements in the system are not those intended, then some potential for compromise may be provided.

(3)   Access Control

This is the ability to provide, on the one hand, and restrict, on the other hand, the connection to a resource. Resources may be databases, other humans, or processing features. Access control is the basis for cryptographic measures that have been in use for centuries. Access control can also be directed at the initiator as well as the recipient of the data transfer.

(4)   Confidentiality

The property of information that prevents it from being made available to unauthorized sources or made available to unauthorized alteration is known as confidentiality. Confidentiality is made available on a connection, interchange, or message basis. Coding and encipherment are the main tools of confidentiality.

(5)   Traffic Flow Security

Traffic flow security provides one or more methods of protecting against traffic analysis by unwanted intruders by providing confidentiality in the traffic links. This method is based upon the premise that considerable information can be gleaned from the knowledge of the message, connection, or interchange traffic alone.

(6)   Connection Integrity (with or without Recovery)

Integrity is the protection of data to insure that it has not been added to, deleted, or modified in any way. An additional feature of data integrity is that it detects any attempts to modify, insert, delete, or replay any of the data. Recovery of the data involves the optional reestablishment of the connection as well as maintaining the integrity of the traffic in question.

(7)   Connection Integrity (Selected Fields)

Integrity of selected fields within a connection-oriented service that embodies formatted transmissions is called Connection Integrity for selected fields. A connection-oriented service is one that requires a setup procedure to establish contact between sender and receiver. Often, these formatted transmissions are broken down into parts suitable for selective manipulation. Such formats consist of fields, one or more of which may be subjected to integrity controls as well.

(8) Connection   Integrity   (Connectionless   and/or   Field Selective)

The other kind of integrity is connectionless, and provides integrity on the basis of a connectionless service offering that may also utilize field selective integrity. A connectionless service, unlike the connection-oriented service, is one that does not require a setup procedure. Only the address of the receiver is needed by the sender in order for the message to reach the receiver and selected fields may be subjected to the process as well.

(9)   Nonrepudiation (Sender)

Non-Repudiation of the sender entails proof that the origin of the data is from the sender. This service provides evidence that the data transmitted did, in fact, come from the sender. This facility prevents the sender from denying that it sent the data in question.

(10) Nonrepudiation (Receiver)

Proof that the destination of the data was at the receiver's end is Non-repudiation of the receiver. This service provides evidence that the data transmitted did, in fact, terminate at the receiver. This facility prevents the receiver from denying that it received the data in question.

Major Functions

Security management has several major functions associated with it that include: (1) data storage protection; (2) data transmission protection; (3)

data receipt control; and, (4) data transmission control. Within these functions are the services discussed above.

(1) Data Storage Protection

The mechanisms associated with this function are mainly confidentiality issues, such as data encryption, access control, and data integrity. Data storage, as used here, refers to the electronic entry, repository, and retrieval of confidential data.

Database storage protection, or database security as it is sometimes called, takes one or all of several forms. First, the data may actually be encrypted in its stored form. This requires an algorithm to convert these data as they are being saved into the database or retrieved from it. Second, varying levels of additional protection may be applied at the field, record or file levels. A field contains information about an attribute, such as eye color or data transmission rate. A record contains a group of attributes associated with a specific entity, such as a person or a piece of transmission hardware. A file contains a set of records that pertain to a group of related functional items, such as human beings or telecommunications equipment items. This protection may require a password for each of these protection stages. Thus, entry into a file may require one password, entry into a particular record may require yet another, and access to a specific field may also need a third.

(2) Data Transmission Protection

Data transmission includes encryption, access control, data flow security, and data integrity. All the elements of data storage protection are also included within this function. Data flow security is a means of convincing a would-be system penetrator that the transmission system is constantly being used so that inferences cannot be made about circuit utilization and, therefore, purpose. Additionally, data transmission protection provides actual means of maintaining data security through the encryption, and access control features of this subject area.

(3) Data Receipt Control

Data receipt control includes authentication, data integrity, and non-repudiation. As indicated above, non-repudiation involves the assurance that either the sender or receiver or both was the one intended as the transmission source or receiving point.

(4) Data Transmission Control

Data transmission control includes nonrepudiation, authentication, and data integrity.

All of these facilities rely upon encryption. Encryption is the altering of data to make it unintelligible to unauthorized interceptors. It comes in two different forms, namely coding and encipherment. Data coding is the substitution of intelligible symbols, phrases, or words for other such symbols that hide the original due to their lack of meaning. Data encipherment is the substitution of letters or strings for unintelligible letters or strings that taken together, convey no meaning at all. The history of encryption closely tracks the history of civilization in that it has been used by all civilized societies since the dawn of man to hide their most closely guarded secrets.

2.3.3   Performance Management

Definition

Performance management is the facility of OSI network management that makes sense of the various reports and provides the system operators with some insight into how the system is performing. It is the means by which "normal" operations are calculated, against which peak and unexpected activity surges are measured, with which, sometimes, faults are detected, with which system changes are assessed, and, quite often, against which the account management function is provided information for system usage tracking. Performance management is the ongoing monitoring and calculation of how the system and its components are accomplishing their assigned tasks, either according to specifications or according to system-defined criteria.

Description

Performance management provides the interface for system "hooks" or termination points that have been placed so as to maximize the collection of operational and health data about the system. The system sensors are the keys to performance measurement. Their integration through the data fusion modules and their interpretation in the decision support modules provide a layered view of the system that enhances the value of performance management.

In operation, performance management collects and analyzes external operational events, and system equipment and functional status. Some of this data is provided by the system itself through the use of event logs. Such repositories are programmed into the system by the designers and implementors of the components making up the system. These repositories provide much needed data that would otherwise be hard to provide, because of the special system software programs needed to extract these data.

The types of external operational events noted by the performance management function are:

(1)   protocols in operation within the system

(2)   errors received

(3)   protocol volumes received and/or sent

(4)   retransmissions requested (errors transmitted)

(5)   special messages received and/or sent, and their disposition

(6)   bottlenecks (time-outs causing retransmissions)

(7)   outages detected (accesses denied)

(8)   categorization of errors

The principal interfaces for performance management are depicted in Figure 2-5. The types of system equipment and functional status monitored and analyzed by the performance management function include:

(1)   equipment and/or circuit status - from fault management (unscheduled)

(2)   equipment and/or circuit maintenance status (scheduled)

(3)   alternate path routing

(4)   situational analyses (time-of-day, major problems, etc.)

(5)   user specific and/or usage group activity profiles

(6)   capability fluctuation analyses (changes in the system configuration and/or capabilities)

(7)   specific client usage (for input to account management)

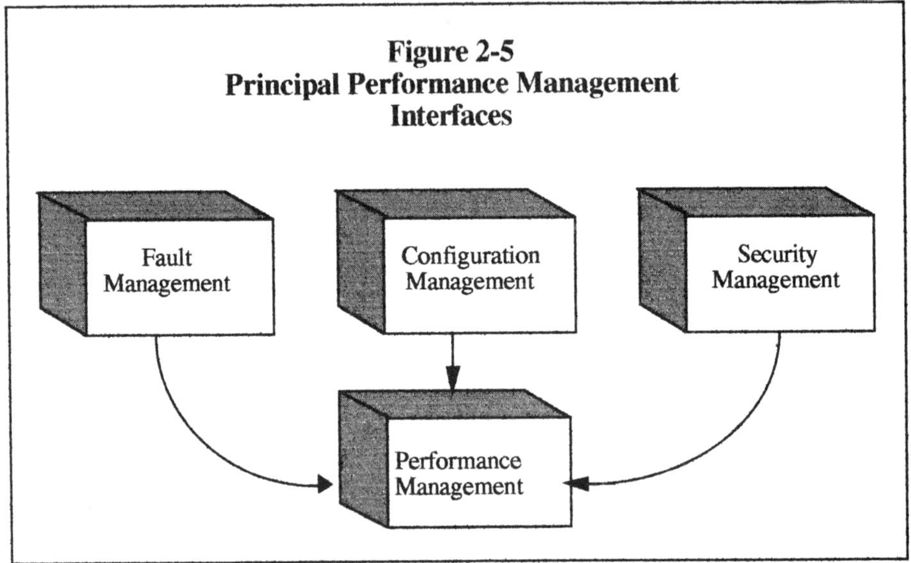

(8)   limits, boundaries, and ranges of operational usage

(9)   analyses of errors, i.e., what and why of errors detected

(10)   event log monitoring

(11)   logged event analyses

Opportunities

Performance management has a wealth of opportunities available to it for smart systems applications. These can be broken down into the categories discussed above, namely external operational, equipment, and functional analyses. The key thrust in most smart systems applications is anticipation or situational prediction. The same applies for the performance management function. Data fusion, decision support, and smart sensor systems technologies could be implemented and have considerable effect over operational status monitoring and analyses, as well as equipment and functional status evaluations, in the following ways:

(1)   **Sensors**
        Sensors are, by definition, monitoring devices that are targeted at specific hardware or processes. The sensors themselves can be either

hardware, software, or both. Because sensors usually act remotely, they can be implemented so as to exercise several types of intelligent features, some of which are expert systems approaches, neural network capabilities, and the like. They can be realized in a variety of ways, some of which are direct attachment to the targeted devices with remote monitoring, software modules unattached to the system, or applications programs making reports via their own messaging facility.

(2) Data Fusion

This is the point at which sensor integration is fused into aggregates of meaningful information. Any abnormalities resulting in errors or other signs of unusual performance, such as retransmissions from time-out conditions, will be integrated in the data fusion center for measurement against steady state conditions collected over long periods of time. Data fusion techniques and technologies can be used to integrate and evaluate operational scenarios, as well as equipment and functional situations.

(3) Decision Support

The focal point for all data fusion and reported status is the decision support station, or executive information system, as it is becoming known. This station, or control point, contains all of the capabilities of interpreting the data being fed into it from the data fusion and sensor systems. The decision support station can be a combination of manual monitoring and subsequent decision making, completely automatic status and control, or some combination in between.

2.3.4   Configuration Management

Definition

Generally speaking, the configuration management function establishes equipment conditions, monitors their status during operations, keeps track of their existence when such resources are not being used, and provides such data as needed to OSI resources so that continuous operations of interconnection services can be maintained.

Description

More specifically, configuration management provides facilities to:

(1)   set the operational and alarm parameters for each resource whether it be hardware or software

(2)   initialize and terminate any of the system resources

(3)   collect data on each of the system components in both a routine state and during periods of significant changes in performance or operational demands

(4)   alter the system configuration as required to meet operational or maintenance demands

Each piece of hardware and software has set points and limits that must be established for operational conditions. The configuration management function maintains a database of and entries for each of these items, for each piece of equipment and software module. Particular interest must also be paid to limits, ranges, and constraints associated with each component so that unexpected conditions can be recognized and dealt with as necessary.

System resources, such as hardware components, can be initialized and closed down remotely from more centralized control points. This initialization process can be activated using command messages to set up the necessary parameters or to initialize a set of internal commands that in turn will initialize the component in question.

The task that makes configuration management fully functional, is its facility to monitor ongoing operations and to continue that process during periods of system stress or unusual loading. This is provided via periodic reports covering two basic issues, i.e., component health and operative conditions.

The key feature of configuration management is its ability to alter the system configuration as necessary to meet operational demands. With the array of multivendor equipment available today, there is no guarantee that reconfiguration can be achieved remotely or even automatically. However, the basic requirement has been satisfied if the centralized control facility is aware of the need for reconfiguration. The method of satisfying this requirement has to be worked out as necessary to meet specific equipment and software needs. These needs can be satisfied in several ways, such as manual reconfiguration with either manual or automatic sensing or automatic reconfiguration with either manual or remote sensing. The key to either method of reconfiguration control is having the information necessary to understand and utilize each equipment type. Many of the key operational features of sophisticated equipment items today have automatic reconfiguration capabilities, so that, for example, data dropouts on a T-1 smart mux will initiate an automatic backup action.

Smart Systems Opportunities

Remote sensing and communications interfaces are important for automated configuration management. If some form of data interface can be established with the various system entities, expert systems can be used to monitor bottlenecks and outages, to perform the necessary reconfiguration tasks, and to predict reconfiguration needs. Software can be remotely monitored and accessed for the reconfiguration of operational parameters.

2.3.5   Accounting Management

Definition

Accounting management involves the tracking of system usage from the standpoint of the users. Usage can be of several types including personal, group, service, or condition oriented, e.g., time of day.

Description

Accounting management requires that the following services be employed:

(1)   an electronic and optional hardcopy listing be provided periodically for all users detailing usage, conditions, and costs associated with such usage

(2)   a directory be established to identify system users, their privileges, and constraints, e.g., financial limits, charge plans, access numbers, and access codes

(3)   incoming usage be sorted and categorized by user so that applicable charging can be made

(4)   incoming usage information be temporarily stored in raw form and storage space be provided for sorted data

(5)   usage indicators be provided to the accounting management function so that storage can be initiated and allocated

    (6)   traffic initiation and termination anywhere in the system provide an indicator of that usage along with appropriate specifications of service type, originator, receiver, length of service, and transmission conditions, e.g., special circuits applied to special usage rates

Smart Systems Opportunities

Expert systems assistants are the primary advantage to be gained from the development of AI capabilities within this area. There are opportunities for monitoring variations of parameters within the accounting function, as well as for predictions for system usage.

2.4 Requirements Definition Process

While it is true that the smart systems implementation of the five functions of the OSI model may employ several varieties of intelligence, it is not always obvious how these varieties are factored into the implementations of these five functions. We will now turn the process of functional implementation around from an OSI-driven perspective to a requirements-driven approach.

    Let's suppose that we are faced with the task of defining and designing a system built around the five functions, but with no other preconceived notions of just how these will be implemented. The first step might be to adopt the model proposed in this book, this being the division of the entire command (management) and control scheme into three principal areas, namely intelligent sensors, data fusion nodes, and finally the decision support module embodied in some sort of workstation configuration. To derive these three areas requires that the requirements must first be defined. There are other methods available to derive technologies, such as functional flow analysis; however, the starting point must be an effective definition of the requirements.

    The generic functional requirements of the network management system contemplated, will be briefly discussed first. The typical network management system needs, at a high level, include: (1) continuous, uninterrupted system operation; (2) maintenance surveillance and problem resolution, including both scheduled and unscheduled maintenance support; (3) connectivity control of system elements and their links; and (4) total awareness of system usage by categories and time (accounting). [8] Broken down further, these top-level requirements translate to the following specifics:

(1)   Continuous Uninterrupted System Operation

    (a)   Fault detection, ambiguity resolution (isolation), trouble reporting, and work order generation, etc.

    (b)   Flow control of communications traffic to accommodate bottlenecks, outages, and peak period overloads.

    (c)   Management visibility into the system operational status for the purpose of enhancing decision support.

    (d)   Provide continuous control of and surveillance over usage of the system so that unauthorized manipulation can be averted.

(2)   Maintenance Surveillance and Problem Resolution

    (a)   Scheduled maintenance program and reporting.

    (b)   Unscheduled maintenance dispatch, diagnosis, parts supply, parts ordering, resolution, and inventory updating, and report generation.

    (c)   Inventory management, to include ordering, status reporting, back order status/control, local supply automatic stocking, failure reports, and specification system.

    (d)   System and component analyses, e.g., reliability, maintainability.

(3)   Connectivity Control

    (a)   Physical and functional linkage compatibility and control, through establishment and maintenance of interfaces through the first 4 layers of the OSI model.

    (b)   Setup and maintenance of user-to-user interaction along the line of the session layer paradigm.

    (c)   Support of the session layer conversation control vis-a-vis presentation and applications layer compatibility.

(4)   Accounting Control

    (a)   Maintain call detail information about all usage within the system.

    (b)   Maintain and modify, as necessary, a database containing

> tariff and/or reimbursement information for cost
> recovery or revenue collections.
>
> (c)   Maintain all records for physical and functional assets,
>   and their maintenance and warranty history.

These requirements can now be allocated to specific functions and hence the technologies can be identified. The allocation of requirements is as follows:

Fault management
Requirements
1(a), 1(b), 2(b)

Configuration management
Requirements
2(a), 2(c), 3(a), 3(c), 4(c)

Performance management
Requirements
1(c), 2(d)

Accounting management
Requirements
4(a), 4(b)

Security management
Requirements
3(b), 1(d)

The final issue now is one of determining how these various tasks will be allocated to the three basic technologies identified previously.

Fault management falls into the smart sensor and data fusion areas of implementation. The detection of operational problems will be divided between sensor technology and functional indicators positioned at the front ends of the system components. In either event, smart sensors, either hardware or software, provide the initial features of the fault management function. The isolation of the specific component, and, in particular, the replaceable unit within that component, will, in many cases, require inputs from other components and a capability for resolving ambiguities. Hence the necessity for data fusion arises.

Configuration management requires all three of the technologies previously identified. The operational condition of equipment items, as indicated by sensor features, provides evidence of usage and connectivity in many cases. The collection of data from several sources, indicating the extent of usage, requires data fusion to be operative. Finally, configuration and/or reconfiguration decisions are best made with as much information as possibly available, and therefore, best done at the decision support level.

Performance management is a data fusion issue where data reports from all over the system are collected, filtered, analyzed, and evaluated.

Accounting management is a combination of data fusion and decision support where systemwide usage data is assembled and summarized at the decision support station(s). Such station(s) will be equipped so as to make predictions and assessments of usage based upon incomplete information available to the system at that point.

Security management is an issue related to decision support. The system administrator presides over the passwords, authentication codes, and other security information necessary to make the system secure and functional.

In the following paragraphs, this basic approach will be explored to yield the allocation of smart systems technologies to these five functions of the OSI model.

Smart Sensors

First of all, the collection of data about the system obviously has to come from system sensors or outside reports obtained mostly from manual sources, such as meter readings. We will address the collection and usage of sensor reports to start with. Keep in mind that smart sensors have three key attributes in common: direct measurement judgment, threshold analysis, and condition prediction.

Sensor reports are *direct measurements* of ongoing system activity. Smart sensors are devices that have the capacity to make judgments about the *quality* and *accuracy* of sensor inputs. They can minimize unnecessary activity and report traffic by comparing baseline data, collected at and by the sensor, to test inputs being collected and evaluated for meaning by the smart sensors. Such evaluations are relayed to the data fusion nodes for further evaluation. Additionally, smart sensors can make their own judgments about certain conditions if supplied with the *threshold* data needed to make such *analyses*. Finally, smart sensors can *predict conditions*, to some extent, as well as the states of incoming monitored values, again, if proper *predictive algorithms* or data points are provided. In some cases, needed data points can be accumulated and updated by the sensors themselves.

Since sensors are so closely associated with fault management, configuration management, and security management, smart sensor technology directly affects the performance of these functions. Attention focused upon specific physical and functional points in the system is vulnerable to smart sensor implementation. Figure 2-6 is a characterization of smart sensor implementation associated with the data collection and analysis processes.

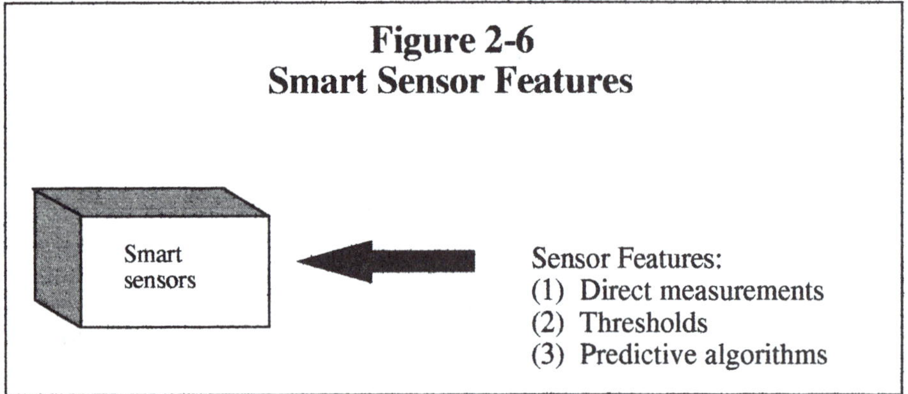

Data Fusion

Sensor information is routed to the data fusion nodes for integration and some degree of differentiation. Differentiation is the process of determining which sensor reports meet certain predefined criteria or priority for importance, which reports are redundant and therefore do not warrant further consideration, or possibly which reports have to be updated based upon more recent information than that available at the sensor. Integration is achieved in several ways, such as by combining, in some meaningful way, the actual number of inputs being integrated at such nodes; another is the identification, categorization, and filtering of these reports. Data fusion is not restricted to the assimilation of similar reports, such as those formatted in the X.25 Protocol or the ATM Protocol, but can be tailored to discriminate between various protocols for the purposes of differential integration with associated intelligence extraction.

Decision Support System

Decision support is the focal point of the operation and is the point at which system responsibility must be exercised, because at that point all available data is present for an informed decision. Decision support is the last stage in the exercise of judgment over the system. All of the five network management functions will, of necessity, be exercised on the decision support workstation.

Summary:

This chapter illustrates how dependent each of the five functions of network management are upon each other. And the application of smart systems makes their interdependence and integration that much closer. The quality of the results and contributions of smart systems directly affect the outputs of each of these five functions. The standards from which the five network management functions come set the requirements for the technologies that can be brought to bear upon their operational utilization. The technologists and system designers charged with the implementation of network management solutions can utilize smart system techniques to provide cost-efficient and rapid outputs from these functions.

References

(1)   Barden, W., *Network Primer*, Microtrend, 1986.

(2)   Bertuol, B., "Sensors as Components for Automotive Systems," *Sensors and Actuators A*, 25-27 (1991) pp 95-102.

(3)   Clair, M. M., "Physical Layer Network Management: The Missing Link," *Telecommunications*, February 1989.

(4)   Ellis, R. L., *Designing Data Networks*, Prentice-Hall, Inc., Englewood Cliffs, NJ, 1986.

(5)   Fleischer, A. G., *Capital Allocation Theory: The Study of Investment Decisions*, Appleton Century Crofts, 1969.

(6)   Goyal, S. K., and R. W. Worrest, "Expert System Applications to Network Management," in *Expert System Applications to Telecommunications*, Liebowitz, J. (ed.), John Wiley, New York, 1988, p 1.

(7)   Green, J. H., *Local Area Networks*, Scott Foresman and Co., New York, 1985.

(8)   McNamara, John E., *Technical Aspects of Data Communications*, Digital Press, Bedford, MA, 1977.

(9)   Sasieni, M., A. Yaspan, and L. Friedman, *Operations Research: Methods and Problems*, Wiley Toppan, New York, 1959.

(10)   Stallings, W., *Handbook of Computer Communications Standards*, Macmillan Publishing Co., New York, 1987.

(11)   Stanley, G., "IBM's TCP/IP effort," *Communications Week*, June 18, 1990, p 6.

(12)   Tannenbaum, A. S., *Computer Networks*, Prentice-Hall, Inc., Englewood Cliffs, NJ, 1981.

(13)   Terplan, K., *Communication Networks Management*, Prentice-Hall, Inc., Englewood Cliffs, NJ, 1987.

(14)   Turner, S., "The Network Manager's Compendium of Standards," *Network World*, April 15, 1991, p 1.

Chapter

3

Sensors and
Sensor Systems

Chapter Highlights:

Sensors and the technologies that support their performance have realized an increase in interest and boost in technological innovations over the past few years. The use of sensors is integral to the achievement of sound product quality, and to the sophisticated operations and functionalities of today's equipment. This chapter will trace the status of sensor technology, the types available, the architectures surrounding their configurations, and new developments in this important area. In the world of digital network management, sensors have taken on new forms, if not new functions that they must perform. The step from analog to digital sensing has left sensor technology playing a catch up game until recent times. The essential truth today is that sensors are peripheral devices that must be as much a part of the system as any other component. Beyond this, sensors are evolving into communications devices, and eventually will be the main control unit, or a backup control unit, for the equipment item itself.

3.1 Introduction

"What are sensors anyway?" This may sound like an unnecessary question, but like so many other terms arising from high tech roots, this term could well apply to any number of devices, processes, or something in between. Basically, however, sensors are mechanisms that provide information about the operational systems with which they are associated, but do not take part in the operation of those systems. [9] Any purposeful, operational system provides the utility and means of accomplishing the user's expressed needs; whereas, the sensor systems provide the health and operational status of these operational systems. As mechanisms for the collection and control of processes, sensors can be either hardware or software entities. Sensors typically provide status measurements of the target system, such as threshold readings, out-of-range reports, and malfunction alarms concerning the system in question. For example, it has been estimated that 15 percent of an automobile's manufacturing cost will go for sensors alone by the year 2000. [1] These sensors will measure such parameters as engine performance, vehicle safety, tire traction, potential dangers to the vehicle, etc.

Sensors are as old as life itself. The first forms of living organisms used sensors to attract themselves to needed surroundings, such as sunlight, or to avoid unsuitable environments, such as land, or water. One of the greatest sensors developed by humans was the flywheel governor that was used to control the performance of electric motors. [12] The governor would control the speed of the motor to which it was attached by use of sensory feedback, that, in turn, would adjust the control of the motor so as to maintain a fixed number of revolutions per minute. Sensor technology has evolved and come into its own with the advent of process-control machinery. More recently, telecommunications has added a new dimension to the need for sensors. Telecommunications now allows sensors to share information with central control points and with other sensors. Sensors have the ability to mitigate information losses, control costs, and protect revenues through their ability to perform remote monitoring of services and circuit conditions. [5]

Another important question to be answered is, "Why are sensors important?" They are important because they provide the link between the network and its control element. As linking mechanisms, they are the source of the collection and conditioning of the network data. The effective control of the system cannot occur without the knowledge of its elements.   Additionally, system faults cannot be detected or isolated effectively without proper sensors. [4]

The use of sensors implies that something can be done about what we are sensing.   This need led to the study of feedback and control mechanisms, and eventually to predictive models used for comparisons of how machinery or processes should work.   The investigations of predictive models spurred the development of artificial intelligence and, in particular, the study of expert systems. [7] Thus, sensor systems and their needs have played a large role in the development of many modern technology thrusts.

3.2 Systems Considerations

The term *sensor* is applied to various methods of information extraction from man-made devices, such as operational telecommunications equipment. This information is obtained and processed in such a way that it is suitable for use as health status, warning, or decision making. These methods are either digital or analog. Analog sensing has been around for many years, and probably originated with cave dwellers as they used sound to signal each other or to monitor the approach of game or hostile tribes. The first examples of digital sensing probably occurred as scouts tried to determine the sizes of opposing armies from the number of camp fires observable at night, or estimate the sizes of animal herds by counting footprints. Most sensor systems in use even today are analog in nature; their collected data is converted into a digital form for use in communications, or stored in some computer system for later analysis.

Typical military acquisition systems today utilize such sensors as seismic, magnetic, acoustic, electromagnetic, laser, infrared, ultraviolet, and visible light sensors to accomplish their missions. [10] These devices are based upon electrical or chemical properties of certain elements or compounds that act as the transducers for the detectable events. Such events are then converted and/or transported to a suitable locale for interpretation and subsequent action. So now, we have the basic elements of a sensor subsystem, namely its transducer, converter, transport mechanism, and interpretation segments.

Smart sensors are varieties of the types of sensors mentioned above that have the ability to adapt to the input information. This adaptation, in turn, provides facilities for intelligent control of the network elements and their modular replacement units.

Sensors consist of several key components that are outlined below and which will be discussed in detail later.

    (1)   sensing element,
    (2)   communications element, and,
    (3)   processing element.

The sensing element may be nothing more than a parallel tap to capture the health status of a piece of telecommunications equipment. In this instance, the telecommunications component implies its status through the operational data carried in its output. On the other hand, the sensor may be a transducer element. The communications portion of the sensor may be a direct link between the equipment and the monitoring station, or it may be much more sophisticated, as with a bus system that interconnects many sensors and supports communications between these sensors and their central control point. The processing element may be a computer fully capable of filtering, predicting, tracking, and adapting to sensory inputs.

There are only a couple of ways that sensor systems can be implemented, i.e., by being on-line with the primary operational system, where the operational components and sensors share their communications links in order to transport sensory data, and secondly, by using their own separate communications links.

3.3 Evolution of Sensor Systems

Sensor systems, for telecommunications, have undergone a consistent evolution of capability. These steps were briefly discussed above and are categorized below to indicate the progression that can be employed in their implementation:

(1) First-Generation Sensor Systems

Analog devices were directly attached to and monitored the hardware for which they were targeted. These devices were connected to the monitoring panels and control consoles. Most such sensors are manually read and evaluated.

(2) Second-Generation Sensor Systems

Analog devices were attached to hardware elements within the monitored network. These devices contained certain features that alerted the remote display consoles of their condition, such as impending failure, accuracy status, etc.

(3) Third-Generation Sensor Systems

Integrated analog devices are located within the confines of the hardware unit featuring not only the transducers and signal conditioning elements, but also an input/output (I/O) element that allows two-way signalling, sensor data in one direction and control in the other. [3] Most current sensor systems are of the third generation variety.

(4) Fourth-Generation Sensor Systems

These systems are comprised of integrated sensor packages containing their own computers. Such systems perform complete processing upon the intended output data prior to its transmission to the control center. This processing may include filtering, conditioning, compression, etc.

(5) Fifth-Generation Sensor Systems

Fifth generation integrated sensor packages calculate and pass compensation coefficients to the central processing unit. [2] The central processing unit then performs its algorithms to calculate the required adjustments and relays commands back to the target unit via the sensor.

(6) Sixth-Generation Sensor Systems

These sensor packages perform their own compensation upon the data stream, and take independent actions to control the input source should that option be authorized by the centralized control facility.

The evolution of the generalized monitoring and control system has seen an emergence of intelligence in the overall system within recent years. Sensor systems, as part of the monitoring and control systems, have seen a gradual transfer to and accumulation of intelligence within these sensor components. This is due in part to the emergence of faster and faster computers, and in part to the sophistication of processing algorithms. Thus, intelligence about the system operation, its health needs, and maintenance issues is slowly being moved to the peripheral components of the system, namely the sensor system.

Not only have sensors evolved as hardware devices, they have evolved into software modules that have special status as applications programs responsible not only for monitoring but also for command and control of the system elements. The next logical step for sensor technology is to evolve software modules that are similar to present-day computer viruses, but acting in such a way as to bury themselves into the hardware and report on their host's status on something like a cycle-stealing basis using the host communications facilities for data transfer. In the next section we will explore the architectural issues of sensor systems.

3.4 Sensor Systems Architectures

The overall system considerations of sensor technology have increasingly focused upon the distribution of the intelligence of the sensory requirements to the locations of data acquisition. The whole reason for discussing sensor architectures is for the purpose of identifying trends in the evolution of intelligent sensors, and to make judgments about the extent to which intelligence can be infused. Figure 3-1 provides an overall orientation of the sensor network. We will now discuss each trend in turn.

3.4.1   Distributed Databases

The distributed database is discussed here because the trends toward such database systems within the realm of general computer networks are likely to be ubiquitous within the near future, and even more likely to be a common attribute of the sophisticated sensor system. As a result, distributed database technology is an important aspect of sensor systems discussions.

The distributed database relies upon certain key features for its operation. These features include the establishment of a "home base" for each data record, the protocol for updating each data record, and the method of communicating data record changes to other interested elements within the distributed environment. These communication methods can either be rendezvous or message passing.

Distributed Database Technology - Data Structures

Aside from the operational methodology associated with the distributed database, it must also have an effective data structure design so that its structure can be implemented in a manner consistent with the operational methodology of the distributed database system. The data structures required for a distributed database include the following data elements:

    (1)   association with the specific equipment item, i.e., the name and/or I.D. number of that equipment
    (2)   association with the specific sub-element of that equipment, such as a major component
    (3)   sensor component identification number
    (4)   nominal range for the point monitored along with its name

(5)   collected values over some period of time, or over some collection interval

(6)   filter, adaptation, or prediction values and coefficients to be used for comparison

(7)   ownership identifier

(8)   copy flag

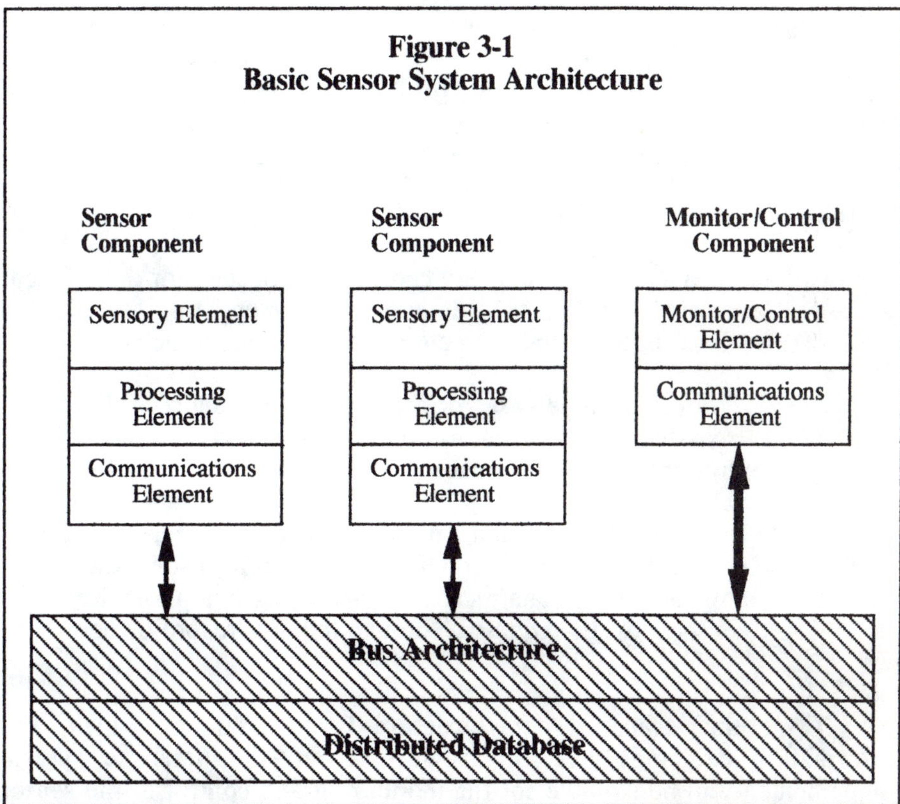

First priority for the update of any of these fields within the data structure should be given to the processor having primary responsibility for that data record as reflected by an ownership identifier (ID), which in this case is synonymous with the sensor component identifier. This identifier can be the address or some coded reference to the specific sensor in question, so that other sensors will know that they are not to update data records which do not match their component identifiers, or IDs. Each data record also is identified with a specific equipment unit, and sub- element within that equipment item.

The remaining data fields within the data structure contain information about sensory norms or operational ranges, collected values over periods of time, and various filter coefficients used for prediction, adaptation, and/or filtering. Adaptation in this context means the adjustment of operational constraints to suit moving or drifting conditions. Clock drift correction is a good example of the use of adaptation algorithms. The copy flag indicates whether this particular data set is a copy or the original data set. Such an indicator is important for the control of the update function, where only the original data set should be updated, otherwise nonsimilar duplicates of the same data record might be passed around the system, and thus cause errors to propogate in the operational environment.

Distributed Database Technology - Operational Rules

The distributed database must embody certain rules needed for the efficient and effective use of the data records and for their updating and sharing among the various sensor users. Some of these rules might include:

(1)   a primary "home base" or owner for each data record.
(2)   "read-only" permission for all fields within a data record by any sensor or other processor other than the owner of that data record.
(3)   any duplicate copy of a data record made and/or used by any other sensor cannot be relayed to any third sensor.
(4)   any update to a data record not belonging to a particular sensor must be requested through the data record owner.

Once the issues of data structure layout, database rules, and information exchange have been worked out, the database system implemented will be suitable for the monitoring and control of the sensor system.

3.4.2   Bus Architectures

The bus architecture is the framework for interconnectivity between the individual sensor components needed to communicate between themselves and their central control component. [14] More specifically, a bus system is a collection of electrical lines that supply the physical components of the system with control signals and data. There are several varieties of bus technologies that may be used to accommodate sensor networks. Each bus architecture contains several layers of functionality and utility. These include the physical, link, and, at least, the network layer.

Numerous bus architectures have been developed over the years to facilitate computer back plane, and computer-to-computer interchange. These include S-100, Multibus, Future Bus, and VME to name a few. Since sensors have progressed to the point that they are essentially specialized computers, an understanding of bus configurations is important in determining their intelligence value.

The trend today is toward "open systems" solutions that provide greater flexibility between equipment types and between equipment units and their sensors. Bus designs incorporate features that facilitate or inhibit communications between various components. These features include bandwidth capabilities, priority levels, addressing, hardware/software identification, and interrupt options. Even though bus structures are not generally thought of as having an intelligent component, they do, in fact, have the ability to reroute messages to other or additional addresses at which their contents may be needed for use. In addition, for those bus architectures where the proximity of sensors to each other is important, communication can be impeded or enhanced accordingly.

Since information dealing with sensor systems is transferred vertically, i.e., between equipment and sensor, as well as horizontally, i.e., between sensors, there is a need for an organized layered arrangement of functions. These layered functions parallel the OSI model of telecommunications, and, for sensor systems, yield two important feature requirements necessary in any effective sensor system. The first is openness, which means that different equipment, from different manufacturers, can be utilized, with sensor systems, in the same network environment. Adherence to the OSI model facilitates the realization of openness. The second feature requirement is transparency, which means that information can be transferred and shared between communicating entities. The information available for sharing can be any services, objects, and models which the systems use to operate.

3.4.3   Communications Interfaces

The sensor communications interface element provides the handshaking and formatting for communications interchanges between the sensors and their bus architectures. These elements operate in conjunction with established protocols to provide the data interchange required. In the case of the OSI model, CMIP (common management interface protocol) may provide the interchange medium between the sensor system and the specific network management system to which the sensors connect. Another popular network management protocol used for the collection and analysis of network element data is SNMP (simple network management protocol). Both of these protocols provide the potential for the fast, effective, and knowledge-based utilization of the communications interface between sensor and central control component, sometimes characterized as a smart system.

3.4.4   Processing Elements

There are innumerable techniques available for use in the analysis of sensory inputs from network elements. Presented below are just a few examples of such techniques.

As mentioned above, the typical smart sensor will have several attributes embodied within it, these being fault resolution, functional redundancy, decorrelation, prediction, adaptation, and compensation. The following paragraphs will discuss these issues within this general category, after which a solution will be proposed.

Fault Resolution

Fault resolution within the system includes 2 principal aspects, these being fault resolution within the sensor or fault resolution within the network elements. The generalized problem within fault detection includes isolation and fault separation. Unresolved fault detection could be approached by defining a model of the system a priori so that system responses could be tracked and deviations outside certain thresholds identified as faults. This approach might be characterized in Figure 3-2.

The system model is a representation of the system as best as the modeler can replicate it. The model attempts to replicate the system operation to the extent possible and to flag those situations in which the real life processes deviate from their norms. This is performed in the comparison unit, and is usually a comparison software module. The decision unit provides the opportunity for intelligence or human intervention to control the resulting actions associated with the management of the process. [13] The question, however, is how the system is able to distinguish between sensor faults and equipment faults.

The typical sensor message consists of either a purely digital response that is conveyed to the network management system, or consists initially of an analog response that is converted to a digital signal and then conveyed to the network manager. [6] Both sensors and equipment elements may experience either hard or soft failures.

Hard failures are sudden failures of network element (NE) functional performance, while soft failures are gradual deteriorations in the functional performance of the NE. Hard failure indications that result in continuing functional performance of the unit under scrutiny are indicative of sensor failures, whereas disruptions in the functional performance of the network element are indicative of an NE failure. Soft failures that result in continuing operations without operational degradation over a period of time can likely be traced to the sensors. Soft failure indications that result in some degradation of performance over time, are indicative of NE problems.

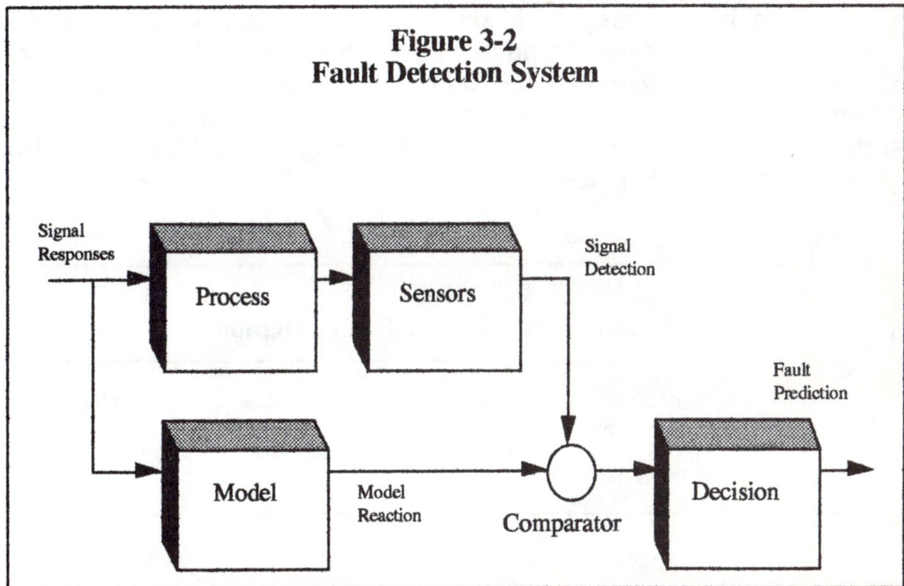

Figure 3-2
Fault Detection System

Functional Redundancy

Whether the situation is one of soft failure or hard failure, the characteristics of the conditions representing potential faults can be used to deduce the nature of the fault, in most cases by drawing inferences from the conditions manifested. Shown in Table 3-1 are simplified conditions that illustrate the analytical process involved in the resolution of system faults. For those situations where there is unresolved ambiguity, the responses from two or more sensors are needed to provide what is termed *functional redundancy*. Functional redundancy provides consistency checks between sensors so that sensor problems can be ruled out over actual equipment problems.

Sometimes the sensor may logically represent the performance of two or more NEs or major systems within one NE, or it may represent the interface between two NEs. In such cases, the ambiguity among the choices must be resolved in order to carry out diagnostic and/or decision processes as to the route of action. This resolution in many cases has to be inferential, utilizing knowledge of adjacent sensors, the nature of the current performance, and recent history to determine which component is to blame, or whether the sensor or NE is at fault.

Decorrelation Technique

One approach to solving such ambiguities is to use what is called a *decorrelation technique*, or prediction method. [8] The decorrelation technique seeks to segregate the failures from the sensor reports so that either one or the other can be pinpointed as the source of fault. This technique consists of three main tests, these being the hard failure test, bad data test, and soft failure test.

Table 3-1
System Versus Sensor Fault Isolation

	Hard Failure Indication/ Functional Discontinuity	Hard Failure Indication/ Functional Continuity	Soft Failure Indication/ Functional Discontinuity	Soft Failure Indication/ Functional Continuity
Sensor		X		X
Equipment Element	X		X	X

The hard failure test is relatively simple and straightforward. It is determined by answering the question, "Does the sensor continuously exceed some threshold or range value previously set for it?" If the answer is yes, then adjacent sensors are searched for corroboration, and upon receiving no evidence to the contrary, the assumption is that a sudden failure occurred. If several failure reports arrive at the central control point almost simultaneously, then another type of ambiguity has to be resolved, namely, which reports represent the actual outage, and which ones represent NEs that are affected by the outage.

The bad data test assumes that intermittent error reports can be resolved by successive data collection to rule out statistical errors. Historical information is retained to use periodically for assessing the likelihood of bad data being collected. This data forms the basis for a statistical distribution from which the probabilities may be drawn to assess the chances for so many bad data reports within such and such times.

The third major type of error is the soft failure, characterized by gradual performance degradations by one or more elements within the system. Gradual component failures can be modeled using decay functions.

Once these three tests are conducted, an estimation of the system condition is derived and the consequent decision is declared. The soft failure mentioned above is a particularly difficult issue with which to deal. A soft failure is one that occurs over a period of time and is very gradual in nature. A classic example of a soft failure is the deterioration of a satellite battery system, or any battery system for that matter. After launch, the batteries for a low earth orbiting satellite, without a solar recharging system, may experience some rapid drop off in voltage levels, say from about 28 volts to about 26 volts. At this point further deterioration is manifested in minimal amounts, down to about 24 or 25 volts, after which the battery begins to fall off again at a steeper rate until it is depleted and unusable. The period between 26 and 24 to 25 volts is a long drift of minute changes hardly perceptible and subjected to considerable noise. The outward impression is one of a flat response to consumption demands, but the actual situation is one of gradual decay.

This gradual decay can be modeled, often easily, with first or second order equations fitted to the data collected from the sensors. The kinds of sensors that we are discussing could be used to collect, fit, compare, and predict failure points as well as times of failure. The second order equation general form is chosen for this illustration, and is given as follows:

$$p(t) = a_0 + a_1 t + a_2 t^2$$

where:

$p(t)$	=	value of the function at any point in time, e.g., battery voltage levels
$a_{(0,1,2)}$	=	coefficients of decay, i.e., representations of wear and tear on the battery system
t	=	time increments of the analysis process

The coefficients that replicate this error trend are found as a solution of three equations with three unknowns as shown below:

$$a_0 \quad\quad + a_1\sum t_i + a_2\sum t_i^2 \quad\quad = \sum x(t_i)$$
$$a_0\sum t_i + a_1\sum t_i^2 + a_2\sum t_i^3 \quad\quad = \sum t_i x(t_i)$$
$$a_0\sum t_i^2 + a_1\sum t_i^3 + a_2\sum t_i^4 \quad\quad = \sum t_i^2 x(t_i)$$

The coefficients are indicative of the magnitude of the system degradation or soft failure involved. If the coefficients are small, the trend is slight; if they are large in value, the trend is drastic in effect. This type of analysis can be used to filter out the effects of the performance decay so that random errors can be assessed as well.

The concept of performance decay is illustrated below in Figure 3-3.

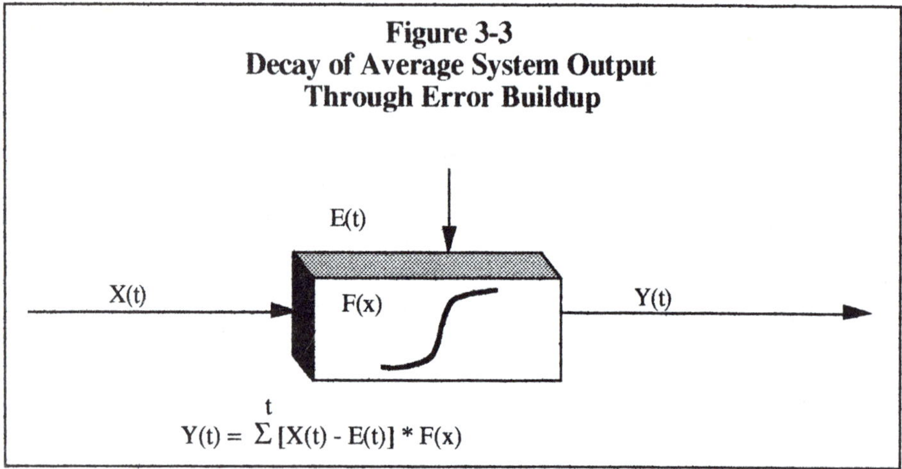

The general concept is for the error E(t) to influence the NE operation, where the normal operation is for the input to the system, X(t), to be influenced by the system transfer function, F(t), such as [X(t)*F(t)], over some period of time. In such a situation, the output Y(t) is affected as a gradual decay, since an increasing error, diminishes the overall system input, and thus causes a decay of the average output.[11] This is referred to as soft sensor failure.

A visual representation of the process of performance decay is shown in Figure 3-4 where the NE performance rate varies as a function of time or events processed, which is approximately the same thing. The long term trend shows a gradual degradation of performance, i.e., increased error rate, while the short term effects show a considerable variation in performance that mask the long-term effects. A second-order function models this type of deterioration very well.

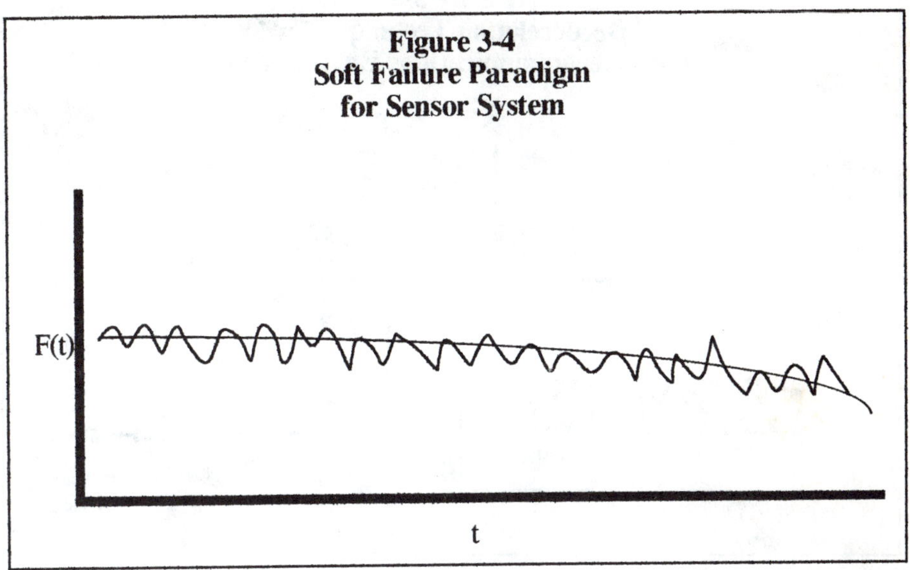

The *decorrelation technique* now can be applied to the problem at hand. The application consists of setting up a filter that compares simulated or model faults with actual sensor data to determine if simulated faults match or mismatch sensory inputs. If sensor data causes mismatches through the filter with what the model provides in terms of faults, the sensor is designated as the source of error. If there are no mismatches between sensor data and model simulated faults, the NE is said to be at fault, if an error report is presented. An example of the decorrelation method is illustrated in Figure 3-5.

Predictive Algorithms

Another category of sensor data analysis is that belonging to the area of prediction. Prediction implies that future parameter variations will be similar to those that have existed in the past, and therefore, capable of being modeled. If this is not true or partially not true, then we are dealing with uncertainty. Uncertainty is primarily an area associated with probabilities. Therefore, modeling and uncertainty are the key topics within the area of predictive analysis.

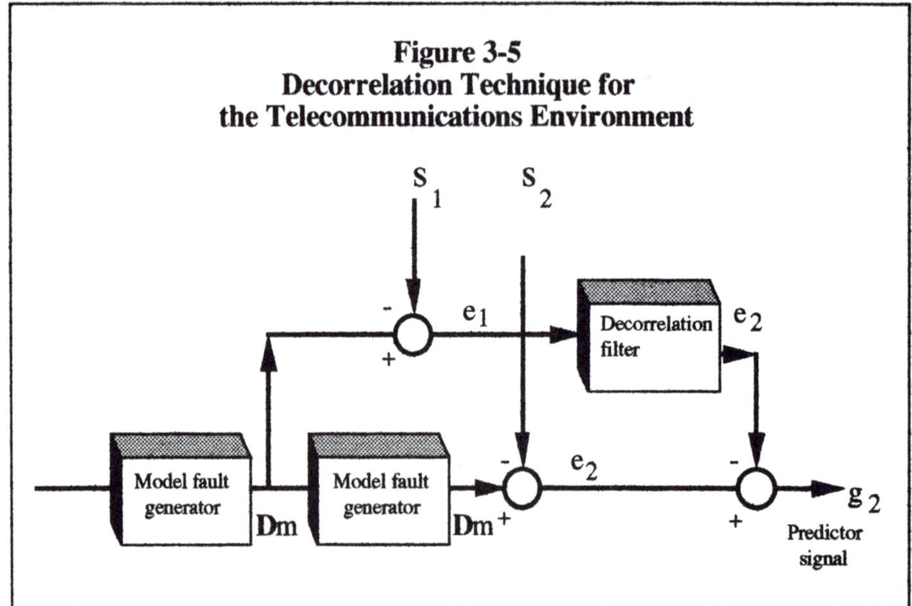

Adaptive Algorithms

One of the issues associated with smart sensors is that of adaptation. Adaptation is the process of adjusting the sensor response to accommodate slow or rapid, irregular deviations within the signal stream. In other words, it is, in effect, tracking the signal, or it is the adjustment of set points based upon the weighting of events. An adaptive algorithm is one in which estimates of the signal parameters are derived based upon collateral information. This collateral information is supplemented by actual observations to yield the weighted response of the signal in question. As time elapses, the direct data reports are given more weight, while the collateral data becomes less of an influence in the weighting scheme.

This type of adaptive problem is solved by weighting the data to reposition the expected set points or expected values of the aggregate data set. The process is tantamount to using indirect information to rate and gauge activity until direct measurements are available. The estimate of the parameter in question is given by the formula:

$$E = w * \chi + (1 - w) * \mu$$

where: w = weighting value
         χ = mean of the direct measurements

$$\mu \quad = \quad \text{mean of the collateral measurements}$$
$$E \quad = \quad \text{estimated mean of the data set}$$

Thus, if the weighting factor is increased, the influence of the direct measurements is increased; if the weighting factor is decreased, the collateral data has the greater influence. The weighting factor is derived from information about the errors in the observed data, i.e., the squared deviations from the mean, or standard deviations. The calculation for a weighting factor is:

$$w \;=\; \frac{n}{n + \sigma^2 \big/ \sigma_\mu^2}$$

Where:

$$w \quad = \text{weighting factor}$$
$$n \quad = \text{weighting category, e.g., time in years}$$
$$\sigma^2 \quad = \text{squared deviation for the observed data}$$
$$\sigma_\mu^2 \quad = \text{squared deviation for the collateral data}$$

The weighting category (n) is the independent variable that conveys the units of comparison, such as units of time, weight, volume, or area. Initially, the weighting factor is highly influenced by the weighting category; for instance, if the weighting category (n) = 4 and the ratio of deviations (σ^2/σ_μ^2) = 2, the weighting factor (w) = 0.67. If n = 8, and $\sigma^2/\sigma_\mu^2 = 2$, w = 0.8. As the weighting factor increases, more credence is given the direct measurements as opposed to the collateral measurements.

Sensor Compensation

Among the problems falling into this category of sensor control are drift tracking, linearization, cross sensitivity, and various response characterizations. Sensor compensation can be applied either at the front end of the system or at the back end. Front-end compensation is performed by exercising a priori algorithms necessary to make the appropriate corrections. Back-end compensation is used in smart sensors, and consists of a correction to be applied to the direct measurement. Since the correction process may vary between sensors, there are no set formulas to follow. However, some principles to keep in mind are:

(1)   the compensation process must be definable by a series of numbers used as the coefficients

(2)   there must be a way of controlling the target variable and the interference variables during the calibration cycle

(3)   the compensation of the target variable must be performed by a function whose response is linear, or if non-linear, the function of this non-linearity must be known

With regard to this last point, it is always critical to be able to determine or predict the regions of linearity associated with a target sensor so that correct compensation can be applied. Linear compensation requires some known relationship between the variables involved. This relationship has inherent errors associated with it that are defined by the derivative of the function defining that relationship. The first derivative provides a measure of the error in y for fluctuations in x. Since the first derivative is an inflection point and one for which the slope of the function is 0, changes in x yield no appreciable variations in y. Thus, the error is 0 at such points. For instance:

for the function,	y	$=$	$f(x)$
the derivative is,	y'	$=$	$f'(x)$
or,	dy	$=$	$f(x)\ dx$
If,	y	$=$	$2x^2$
then,	dy	$=$	$4x\ dx$
or	dy/dx	$=$	$4x$

Therefore, an error of 0.1 in the measurement of x becomes an error of 0.4 in the calculation of y.

Nonlinearity

For non-linear situations, the region(s) of nonlinearity may be discontinuous, thus requiring two or more functions to describe the regions of interest. Figure 3-6 illustrates the issue associated with a nonlinear function. Note the discontinuity in the function portrayed. While a polynomial may describe the relationship between x and $f(x)$, a discontinuity in the function may require more than one set of relationships to resolve values of $f(x)$.

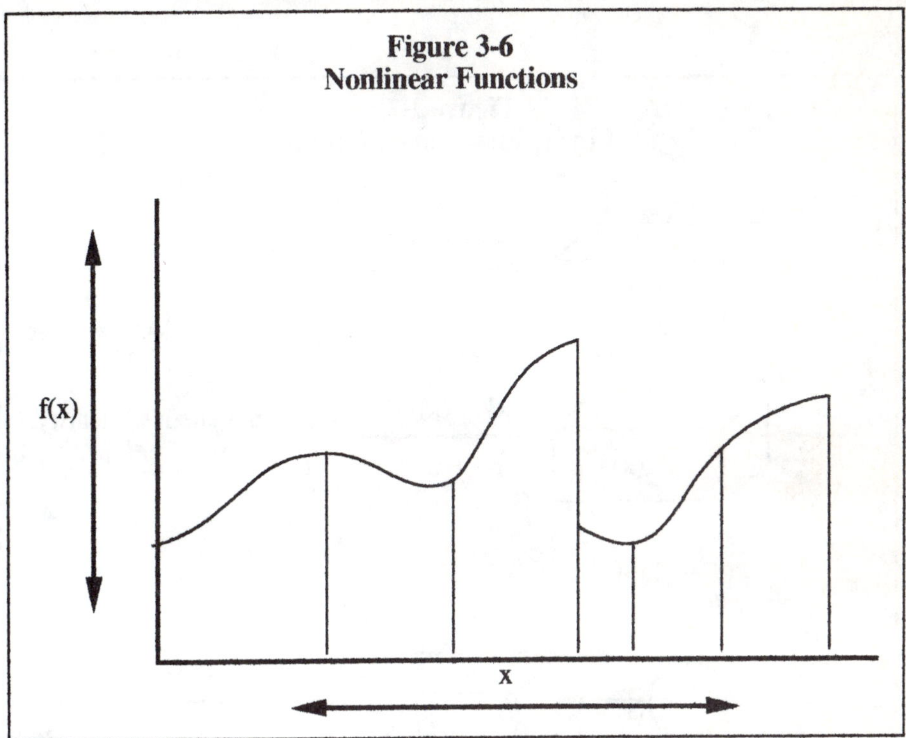

Figure 3-6
Nonlinear Functions

The question of linear or nonlinear can be defined as follows. A linear function is one in which there is an incremental change in y for each change in x. If this relationship holds true for one segment of the function, but not for other segments, then the system has only a partial linear operating range. Therefore, a nonlinear system is one in which incremental changes in x yield variable changes in y. It is important to be able to classify systems as being linear or nonlinear across either all or some part of their operating range. Most signal response study is oriented toward analyses of linear systems rather than nonlinear systems. Unfortunately, most signal sources have nonlinear components and/or responses and therefore it is important to understand and analyze the characteristics of nonlinear systems.

The tests for nonlinearity consist of additivity and homogeneity. Homogeneity is the property of a function such that its output is the same whether the input is multiplied by a constant before or while being subjected to the transformation process. Linearity is the property of a function such that its output is the same no matter whether its inputs are added prior to the transformation or transformed separately and then added. Figure 3-7 illustrates these concepts.

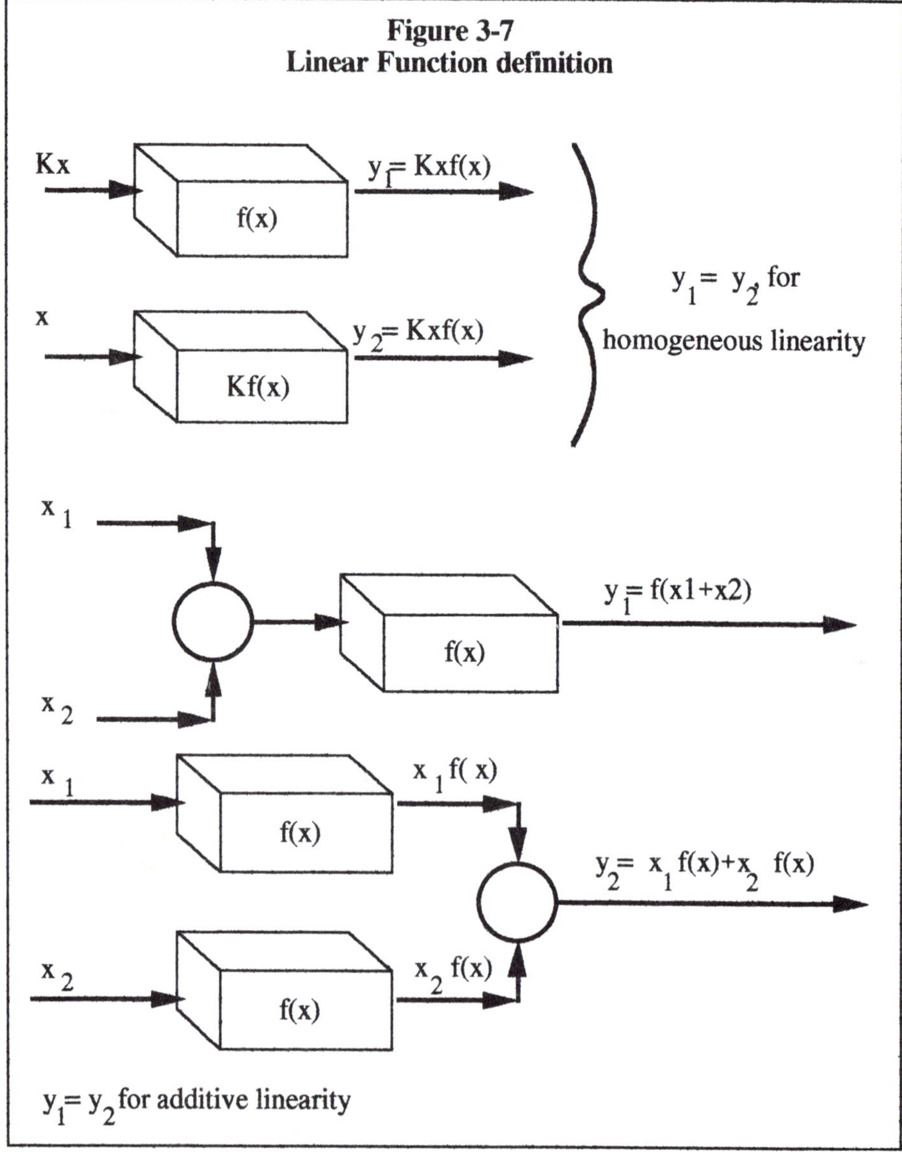

Signal Correlation

One of the methods of determining status or set point of an expected signal is to compare it with some known signal pattern. This is known as correlation. The process of correlation is also important in comparing two signals coming from different sources or from different circumstances. The solution to any type of signal processing problem is accomplished by either analyzing the frequency or the time components of the specific signal of interest. Correlation is a process of comparing signals in the presence of noise in the time domain. This process is initiated by first dividing the input signal into time slices. Next, the reference signal, either the input signal again, as in the case of autocorrelation, or the reference signal, as in the case of cross correlation, will also be divided into time slices. At $t = 0$, these two signals will now be subjected to a set of mathematical processes where they are multiplied and added to each other thus compiling a set of statistics from which results about the relative similarity or dissimilarity of the two signals can be drawn.

At $t = 1$, one of the signals is time-shifted one slot, and the cross multiplication of each corresponding time slot in each of the signals is performed again, and these values are collected again. This process is repeated again and again from $t = 0$ through $t = $ n, until the entire set of statistics is obtained. The general form of the correlation equation is given as follows:

Autocorrelation

$$\text{Product} = \sum_{t=0}^{n} \left(V_{(t+1)_1} * V_{(t_2)} \right)$$

Cross correlation

$$\text{Product} = \sum_{t=0}^{n} \left(G_{(t+1)_1} * H_{(t_2)} \right)$$

Convolution

$$\text{Product} = \sum_{t=0}^{n} \left(V_{(t-1)_1} * V_{(t_2)} \right)$$

where: G(t), H(t), V(t) are segmented signals.

Product, in this case, means that time slice products are obtained, added, and accumulated to achieve an integrated picture of the two signals compared.

Thus, the products of each multiplication step are saved and summed over the period of the signal analysis. The resultant function should represent a decaying sine wave for those situations where a strong correlation exists between signals or between two noisy versions of the same signal. The result of this process is a set of values for each time slice that can be plotted. The plot of such a function provides an indication of the relative correlation or lack of correlation between one signal and another or the same signal as viewed through a noisy environment. Figure 3-8 shows a typical correlation function for which there is a strong correlation. This particular function depicts an autocorrelation of two cycles of a sine wave that is 3 dB down in random noise.

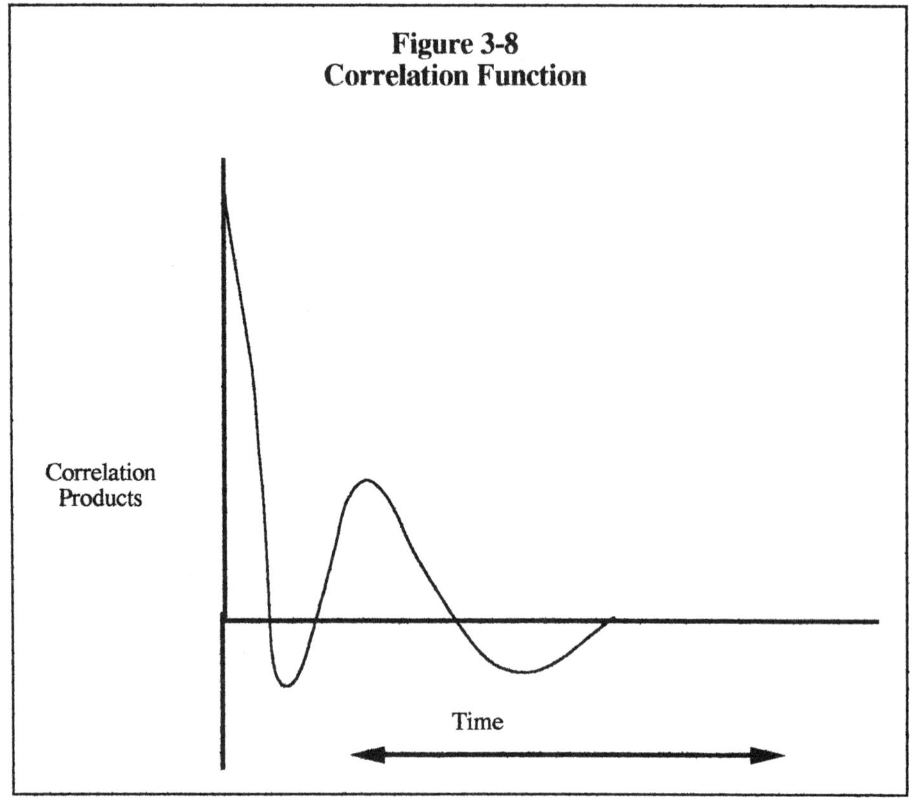

3.4.5   Sensory Element Architecture

For most state-of-the-art telecommunications systems, the sensor is digital in nature. [15] It may receive direct digital reports from its interfacing functional components, or it may receive analog reports and convert them to digital reports prior to the relay of these reports to the report collection point. The key to sensory processing is to provide as few sensors as possible, consistent with the monitoring requirements. These requirements should address the following issues: (1) the number of sensors necessary for a particular situation, (2) the operational parameters to be monitored, and (3) the overlap and redundancy necessary to monitor all relevant parameters. These issues each have their own solutions depending upon the circumstances at hand.

Data Conversion

The first architectural issue to be addressed in sensor architecture is that of data conversion, because the conversion is key to correct collection and interpretation. Normally, data conversion is performed using a linear relationship between the input signal and the digital representation of that signal. The typical conversion paradigm is illustrated in Figure 3-9 and traces the relationship between analog and digital signals.

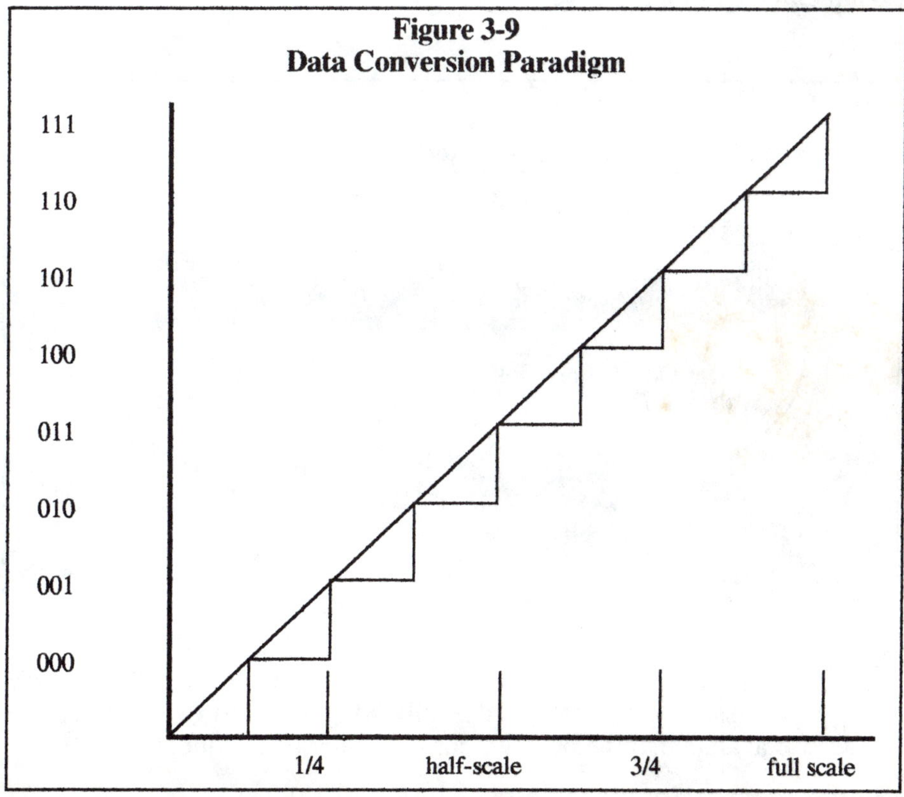

If the first eighth of the input signal is presented to the sample and hold digital conversion circuitry, the output is a digital 000. If the second eighth of the signal is presented, the output is a 001. If the full scale deflection is presented to the input, the digital output is 111. For the illustration presented, the entire analog input is divided into 8 segments, which can be represented by a 3-bit output signal (2^3). If the input scale were divided into 32 segments, the output code would have to be five digits (2^5) long.   The larger the code, the finer the granularity of the segmentation, and therefore, the better the resolution of the process.

Sensor Configurations

It is always instructional to get some sort of feel for the technical issues associated with sensor architecture in order to appreciate the problems that may have to be tackled by those involved with the design, engineering, and/or maintenance of the specific equipment targeted.   As a simple example, refer to Figure 3-10.

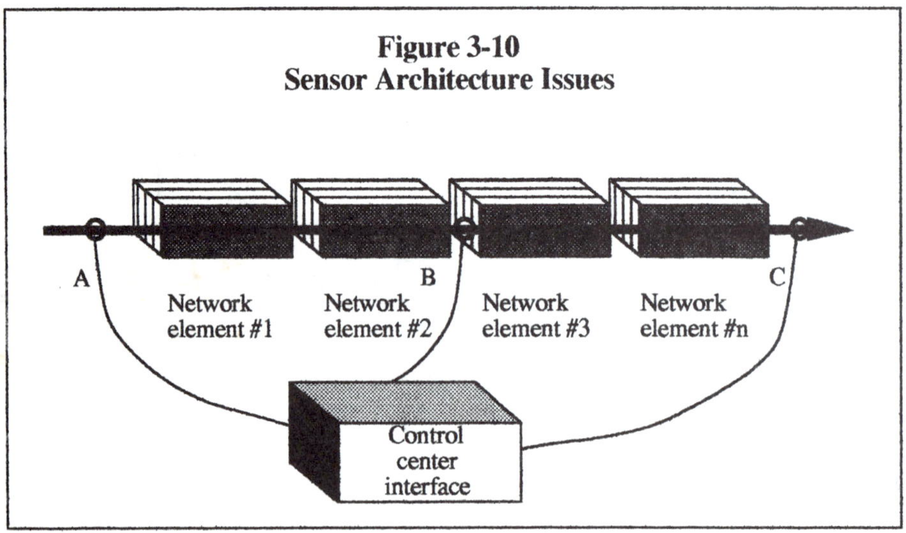

In this example, there are three sensors attached to the linear array of four processes as depicted. The following breakdown indicates the knowledge available to the observer for the conditions defined:

Condition [Operative Sensor(s)]	Sensor Condition [Suspect Units]
C	All four plus input signal
A, C	All four, if input signal OK
A, B, C	Units 1 & 2 or 3 & 4, if input signal OK

Thus, the level of ambiguity suffered even for this simple example is significant if there is no further information with which to deal. Unless there is collateral information that would lead the investigation to the specific offending unit or units, the level of identification is dependent upon the number of sensors and the level of ambiguity tolerated by the system. For a linear, parallel, or ring configuration of operational units, the number of sensors needed to resolve errors is a function of the number of units involved and the level of ambiguity to be born as given below:

$$S_n = U_n / A_l + 1$$

Where:

$$
\begin{aligned}
S_n &= \text{number of sensors required} \\
U_n &= \text{number of units involved in the system} \\
A_l &= \text{level of ambiguity to be tolerated} \\
l &= \text{sensor needed to monitor the input signal}
\end{aligned}
$$

Using this formulation, a system of seven units in a ring configuration, where the requirements are to achieve a level of ambiguity of no more than 2, would require 4.5, actually five sensors rounded up, in order to achieve the results specified. Four sensors would require that at least some part of the system would be exposed to at least three units where no sensor information would be available.

The discussion of architecture must also consider the issues of reliability, validity, and availability. [16] Architectural measures taken to mitigate such problems include: (1) redundancy of sensors per monitoring point in order to increase reliability; (2) use of reference signals or other data benchmarks for comparison with collected data in order to assure the collection of valid information; and (3) internal tracking of maintenance requirements in order to alert the control node of inspection needs.

3.5 Influences upon Sensor Functionality

Aside from the various techniques available to analyze the data collected and conditioned by sensors for eventual transmission to their central control units, sensors are also influenced by external forces and circumstances that affect the quality of their operations. These external forces may degrade the performance and accuracy of these sensors. The design problem then is to weigh the likelihood of suffering the worst case as opposed to an expected situation that is something less than the worst case. Some of the more important influences are discussed here by categories indicated below.

3.5.1   Sensory Overload

Sensors may suffer from at least two important types of overload, saturation and input speed. A system in which the rate of data influx exceeds the ability of the sensor or sensor system to detect and deal with such data, is almost as bad as having no sensory capability to begin with. This situation is known as sensory overload. System designs start with requirements. These requirements anticipate the rate and volume of influx based upon estimates made by the engineering design team of the sensor's ability to cope with the input rate. Two methods are used to handle this type of input rate, one being the speed of the circuitry to deal with these inputs, and the other being a buffer used to handle bursts of input over a short period of time. The speed of the circuitry must also be able to handle the fastest reports of the input data stream.

3.5.2   Detection Classification or Ranges

Another potential problem for sensors is their ability to classify the report as to origin, or location at that origin. Additionally, the scope of the system to detect an error at all is important. A report must be capable of being classified and associated with a particular location and equipment item. Many times a single sensor cannot make a positive detection identification. A positive ID must be made in conjunction with two or more sensors acting in cooperation with each other. Likewise, the number of sensors is all important in the determination of the types of problems as well as the granularity of their evaluation.

3.5.3   Masking

In some situations the report becomes confused, because many other reports are also coming in almost simultaneously. This is called masking. Large volumes of sensory inputs all arriving at the same time, must be sorted out using appropriate criteria for the situation at hand. The important issue is to have such a circumstance planned ahead of time so that these competing inputs can be categorized and resolved as efficiently as possible.

3.6 Sensor System Requirements

Sensors and their systems operate under very demanding conditions. These conditions are dictated by the requirements of the operations involved. The sensors must satisfy the following basic requirements:

(1)   They must be capable of the rapid conversion of data from analog form to digital or vice versa.

(2)   They must have the capacity to store certain data in their own internal databases, and to update that data on download command.

(3)   They must have the ability to perform comparisons with limits, ranges, and thresholds.

(4)   They must have the ability to receive commands that alter their processing sequences or to alter their database entries.

(5)   They must be able to perform continuous calibrations of their internal system based upon collected data either from the operational system or the central control unit.

3.7 Basic System Task Characteristics

The rudimentary tasks that must be performed in a telecommunications situation are depicted in Figure 3-11. These tasks apply, for the most part,

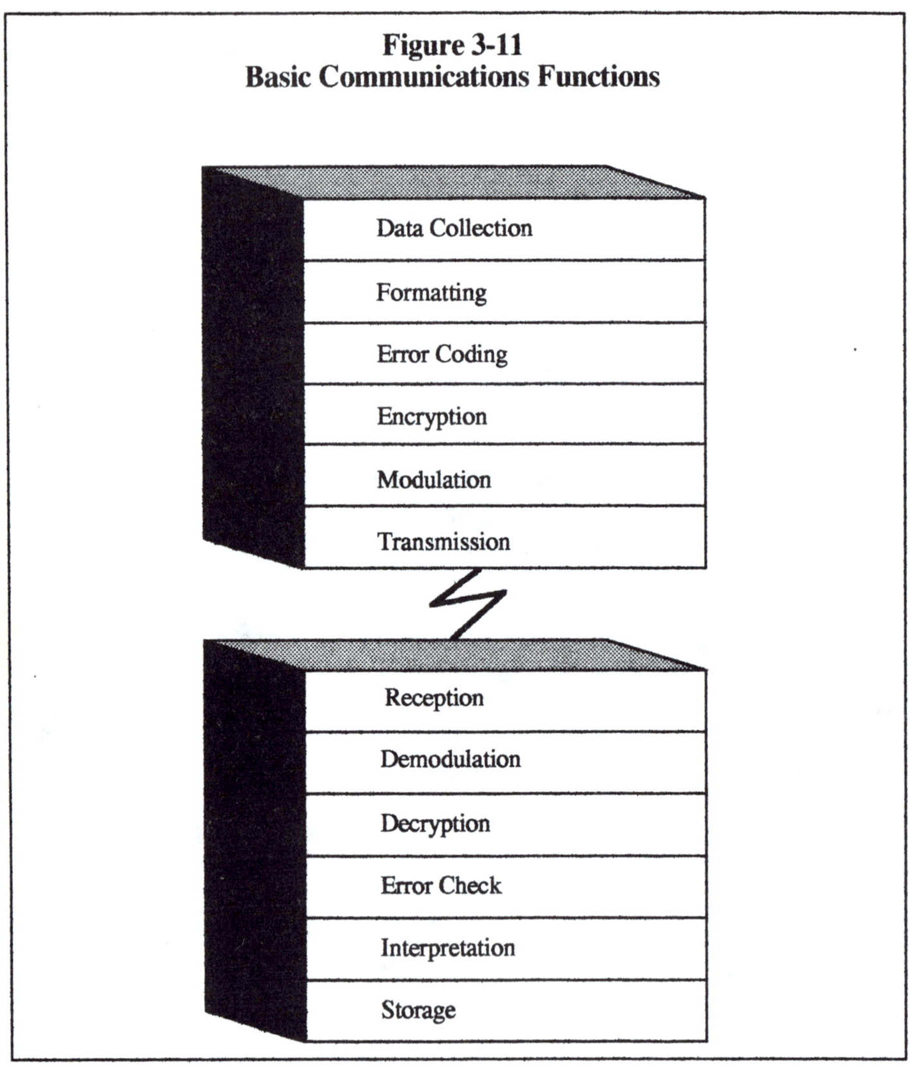

whether the transmission is voice, print, or video, i.e., any type of
multimedia transmission. Conceptually, the network management for such
a system is straightforward, in that each of these tasks must be monitored,
measured, and actions taken where necessary to correct errors or to
alleviate operational problems. The remaining chapters in this book will
endeavor to explore the various aspects of these tasks from the standpoint

of the intelligence necessary to prevent or solve problems relating to these basic activities.

Summary:

This chapter traces the evolution and state-of-the-art development of sensors. The trend for sensor systems is for them to evolve into parallel operational systems whose mission will be to detect, report upon, and control the systems which they shadow. Sensors themselves will evolve into fully functional, self-adapting computer systems. Such sensor units will take on the form of both hardware and software-based entities. Sensor systems will be thought of as redundant systems at some point with almost the same control capability as the primary systems. Sensors are taking on the roles of sensing, evaluation, control and self-correction. Each stage of evolution will increase system quality, performance, efficiency, and reliability for the target system.

References

(1)   Bertuol, B., "Sensors as Key Components for Automotive Systems," *Sensors and Actuators* A, 25-27 (1991), pp 95-102.

(2)   Brignell, J.E., "Software Techniques for Sensor Compensation," *Sensors and Actuators* A, 25-27 (1991) pp 29-35.

(3)   Chen, K., et al., "PASIC: A Processor-A/D converter-Sensor Integrated Circuit," *1990 IEEE International Symposium on Circuits and Systems*, 1990.

(4)   Cortner, J. Max, *Digital Test Engineering*, John Wiley & Sons, New York, 1987, pp 1-27.

(5)   Coughlin, V., *Telecommunications Equipment Fundamentals and Network Structures*, Van Nostrand Reinhold Company, New York, 1984.

(6)   Henning, W., "Bus Systems," *Sensors and Actuators* A, 25-27 (1991) pp 109-113.

(7)   Fischer, M. A., and O. Firschein, *Intelligence: The Eye, the Brain, and the Computer*, Addison Wesley, Menlo Park, CA, 1987.

(8)   Kroschel, K., and A. Wernz, "Sensor Fault Detection and Localization Using Decorrelation Methods," *Sensors and Actuators* A, 25-27 (1991) pp 43-50.

(9)   Lerner, A. Y., *Fundamentals of Cybernetics*, Plenum Press, New York, 1975.

(10) Maxfield, M., A. Callahan, and L. J. Fogel (eds.), *Biophysics and Cybernetic Systems*, Spartan Books, Washington DC, 1965.

(11) Pask, G., *The Cybernetics of Human Learning and Performance*, Hutchinson Educational, New York, 1975.

(12) Porter, A., *Cybernetics Simplified*, Barnes & Noble, New York, 1969, p 23.

(13) Robinson, H. W., and D. E. Knight (eds.), *Cybernetics, Artificial Intelligence, and Ecology*, Spartan Books, New York, 1972.

(14) Schwaier, A., "Progress in Fieldbus Developments for Measuring and Control Applications," *Sensors and Actuators* A, 25-27 (1991), pp 115-119.

(15) Shapiro, F.B., et al., "A Custom I.C. for Multichannel Telemetry with Digital Sensors," *Frontiers of Engineering and Computing in Health Care*, 1984.

(16)   Wagner, U., "Reliability and Fault Tolerance of Low-cost Multipoint Sensor Interfaces," *Sensors and Actuators A*, 25-27 (1991) pp 73-78.

Chapter

4

Data Fusion Processing Systems

Chapter Highlights:

Smart systems derive much of their intelligence from the combinational effects of the collection activities supporting the front end of the system in operation. Data fusion has a historical basis for its justification, beginning with, or possibly even before, the writings of Bayes. The calculations of probabilities of events, given certain other event occurrences, are well known and understood processes today, yet few systems today take advantage of inferential statistics. The fact that data fusion concepts are derived from such inferential statistics makes their implementation even more rare. This chapter is about the development of fusion concepts and their application to today's sophisticated communications environments. There is no magic here. The basic concepts can only benefit the network system if there are sufficient alternate paths for sensors, or if there are different sensors available from which to build confidence levels.

4.1 Background

Data fusion in one form or another has been an accepted technique for centuries. Even in ancient times, multiple reports about some activity or event were collected, collated or categorized, and evaluated in order to make sense of this activity. *Data fusion* is defined as the integration of information from multiple sources, or time-sequenced from the same source for the purpose of producing the most specific possible view and assessment of a particular entity.

There are two principal types of data fusion, direct and indirect. Direct data fusion occurs when the sensors involved and the fusion process are attached to and performed on the equipment being assessed. For network systems, or subsystems, this means that system or subsystem managers have direct visibility into the operations of the equipment components being tracked, and the collected reports are being fused at the point of the manager component.

Indirect data fusion occurs when the fusion process is physically remote from the sensors servicing the equipment or systems under scrutiny. For example, the network manager may be many miles from any of the elements it oversees, and, therefore, raw data reports must be transmitted to the remote manager for the data fusion processes. The entire fusion process may also occur in widely dispersed locales. There is always a spatial and temporal component to each fusion activity. However, reports may be spaced in time or location, and the extent to which they are separated may indicate how they are interpreted.

Figures 4-1 and 4-2 illustrate the basic concepts of temporal and spatial fusion. [3] In Figure 4-1, the concept is implemented by tracking the sequence of activities and interactivities between operating components. For instance, a pattern of activity associated with cooperating components may yield an indication of suspected reactions by one of the components when all are viewed as an aggregate set. The components preceding the one under investigation may yield a particular pattern of activity that is indicated as an error by the third component. Thus, the error message yielded by the third component betrays an error indication previously indicated by the first two components, which in turn may point to one or both of these components being the culprit of the investigation.

Figure 4-2 is an illustration of spatial data fusion. Here, the various status messages of the system are physically separated from one another so that separate identification can be maintained. [4]

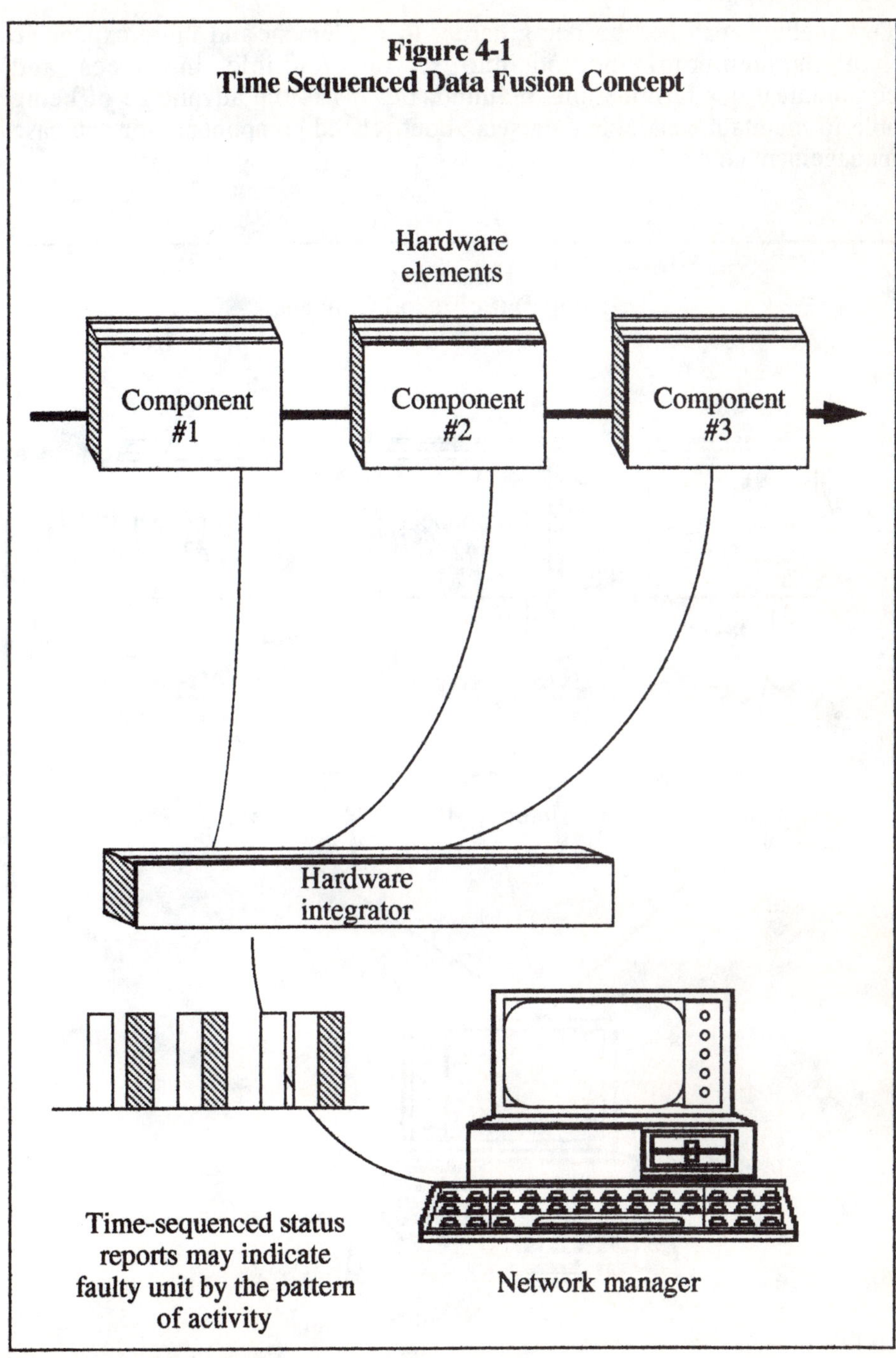

Figure 4-1
Time Sequenced Data Fusion Concept

This method may end up being harder to implement and more expensive than the temporal method due to the multiple interfaces and communications terminations required, but it has the advantage of being able to maintain separable data sets about related components for database management convenience.

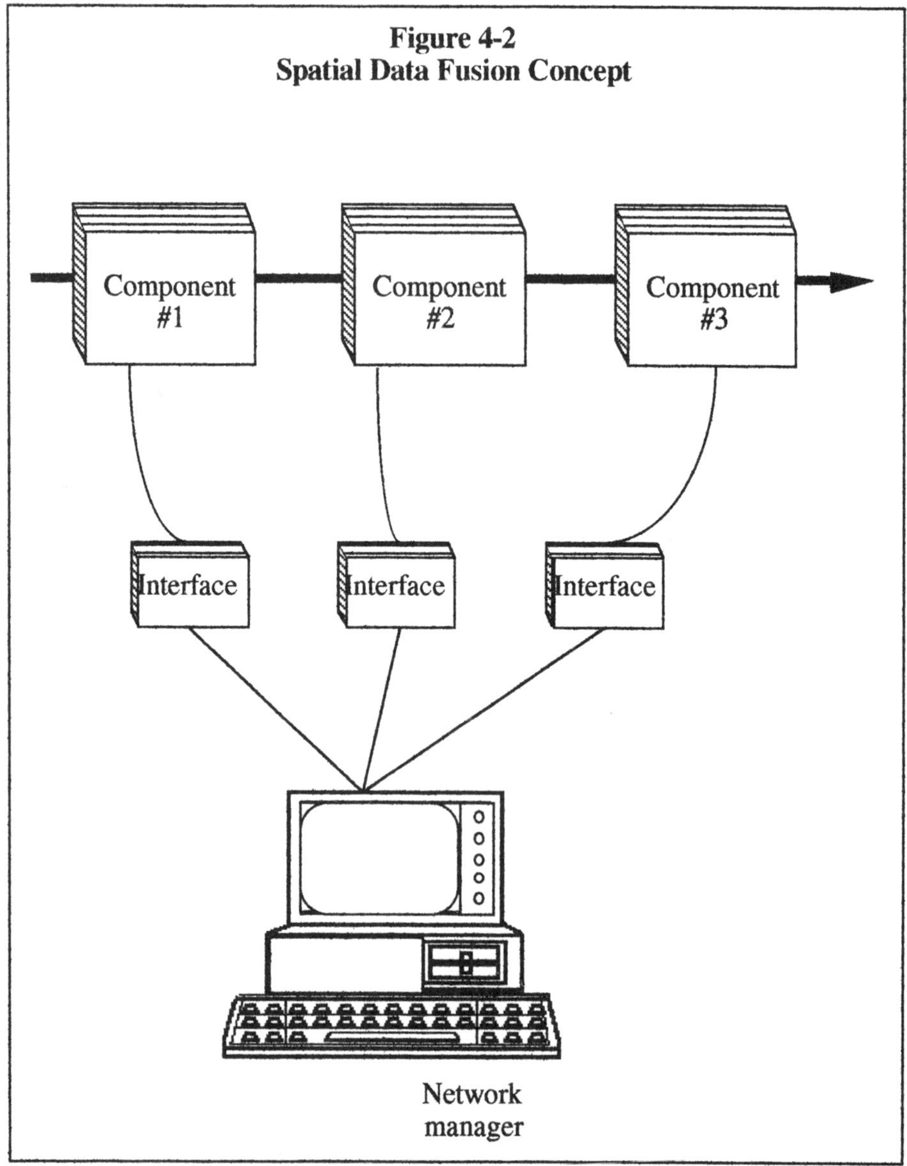

There are six main functions associated with data fusion, these being sensor control, report collection, report correlation, report file storage, report classification, and report evaluation. [8] These categorizations have been developed based upon functional segmentations of the steps through which the process must flow. Each of these functions will be addressed and discussed below.

(1) Sensor Control

Sensors, whether they are locally associated with or remotely located from their fusion process, must be periodically evaluated and appropriately adjusted for biases, drift, etc., that are identified with the data collection process. The control mechanism that deals with these issues has to be accessible to the user and maintainer, and operationally clear.

(2) Report Collection

Reports arrive from many sources and at seemingly simultaneous points in time at the data fusion process control point. Such reports are collected via either in-band or out-of-band signaling systems. In-band control signaling is the method by which the data channel is actually supplying the capacity for the control of data between the source and the control point. Out-of-band signaling is the method of control handling via a separate channel outside the one being used for data. Usually the requests for data are conveyed in the signaling medium. For in-band signaling, the control information is actually carried as a part of the information bandwidth but is usually transparent to the user in order to maintain an organized picture of what is going on.

(3) Report Correlation

Report correlation is the process of making sense of the collected data representing the system sensors. There are several key tasks performed in report correlation, namely deduplication, segmentation, or cluster identification, categorization, hypothesis testing, and probability analysis. Each of these will be identified as follows.

(a) Deduplication

Status reports, especially those generated as a result of faulty operations, may be duplicated several times or in several places. Such reports may indicate numerous problems even though only one problem actually exists. It is the task of the data fusion system to sort out such duplicate indicators and to identify exactly which reports refer to other similar reports and which ones refer to different problems.

(b) Segmentation

This term applies to the task of separating different messages from each other that pertain to different events. In the correlation function, segmentation is an important step in the sorting and analysis activity that must be done in order to resolve issues from nonissues, or related reports from non-related reports.

(c) Categorization

Reports, once received and preprocessed, are subjected to a categorization process that segregates the various unduplicated and segmented reports into like clusters. At this point, the reports are ready for hypothesis testing.

(d) Hypothesis Testing

A series of hypotheses will be tested that relate to possible random occurrences of error, and assignment of errors to other equipment. The probabilities of errors in various circumstances are indicators of problem sources or normal events. In either event, the tests applied help to identify points of network management focus.

(e) Probability Analysis

Probability analysis can best be appreciated by thinking of various reports, coming from many quarters, each with a different level of confidence, i.e, probability of assurance. This task is responsible for trying to separate out similar event sets using their respective probabilities. Thus a string of events with different probabilities may relate to one event or as many events as there are reports. The techniques used for such determinations can vary according to the circumstances involved.

(4) Report File Storage

The mechanisms for storing reports after they have served their initial usefulness are important for retrieval and for making later associations with other reports that will have to be retrieved and compared with their like kind. Subsequent retrievals may be for such purposes as use in artificial intelligence (AI) analyses and decision processes.

(5) Report Classification

Reports, in order to be entirely useful, should be classified according to type, location, timing characteristics, usage designation, etc. The retrieval of information, as mentioned above relative to report file storage, depends upon how it is filed and, in turn, file storage depends upon the type of information required. There are many different ways that reports can be classified simultaneously, and the approach used will dictate the ultimate value of the decisions reached. At this point, the reports left remaining can be referred to as finished reports, because many of the early sets have been eliminated, and the remaining are deemed to be valid and reliable relative to the entire set originally collected.

(6)   Report Evaluation

Using filtered information is part of the decision support function to be discussed in the next chapter, but its utility is enhanced by such tasks as prioritization of the finished products, combinations of some finished reports that may have previously escaped such scrutiny, use of finished reports to generate new hypotheses, and the assignment of confidence levels to those remaining, on the basis of their validity and reliability.

A generalized data fusion architecture that can support all of these requirements is illustrated at a high level in Figure 4-3.

In this architecture, the sensors obtain information from the environment and are assigned to processing elements associated with their activities. The sensors may be of the types described in Chapter 2. The processing elements provide interactive or one-way capabilities that either take the collected raw data, apply certain corrections, and route it to the next stage of utilization, or these processing elements supply the sensors with periodically updated calibration data so that they can make their own sensory corrections.

Message processing is the act of sorting the various messages so that they can be used either for single or multiple sensor analyses, or for sensor management. [9]   Decision support for sensor usage and confidence planning is also a possible destination of the routing function performed by the message processing entity. The message processing function performs its tasks by having some way of identifying the type of message it handles. This identification resides in the appropriate identification field of the message format. The message processing function must decode at least this particular field and then take the appropriate action to route the message to the proper next stage.

Single-source message processing addresses the issue of time sequencing of reports for the purposes of identifying aberrant circumstances or the handling of those circumstances. This timing analysis would usually be performed in conjunction with other events being tracked elsewhere within the system. Conclusions drawn as a result of these analyses are the result of comparisons with known or measured activity, or are the result of conclusions drawn as a result of AI inferences made.

Multiple-source message processing utilizes information from several sensors, either of the same or similar type, that allows the *data fusion* function to take advantage of the spatial dimensions of the inputs in order to make recommendations about system health. Multiple-source messages sometimes provide the system with a higher level of confidence of its status than would ordinarily be the case with single-source messages.

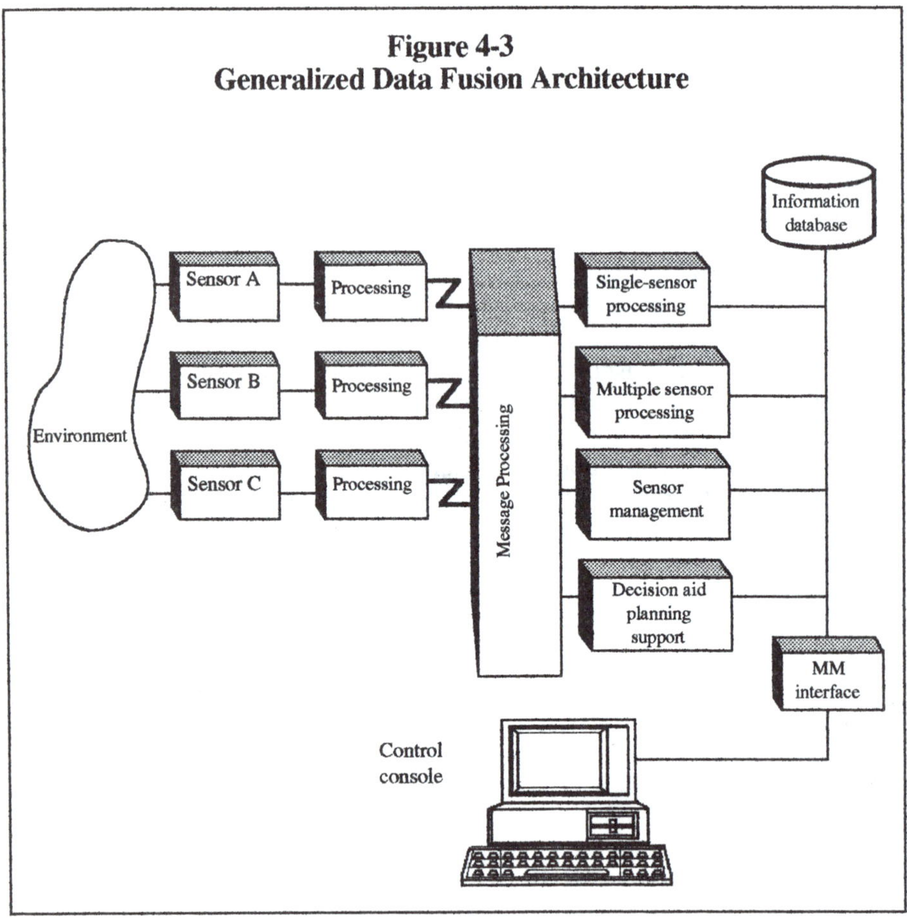

Sensor management is a function that encompasses all of the calibration, alignment, and control necessary for the sensor to operate as independently as possible. It may include such sophisticated features as *goal programming*, where specific objectives are imparted to a specific sensor element, and from which the sensor focuses upon specific conditions that relate to that goal-oriented command.

It may also include a priori models of scenarios that may be encountered, known as *observation modeling*. This term means that specific sets of conditions may be stored into or otherwise made available to the sensor for the purposes of reporting upon such sequences of events as they are encountered.

Additionally, it may provide for *resource optimization.* This is a set of conditions and constraints that can manage the usage of an equipment item. Such constraints might include repetition rates of equipment usage, certain physical limits, etc.

Decision aid/planning support is a high-level capability that requires very sophisticated features and facilities. [8] It is tantamount to having a built-in AI capability. The general features of such a capability are specific planning domains, and the situation assessment of certain narrowly defined results relating to those planning domains.

4.2 Analytical Orientations

We can think of data fusion as being the buildup of a series of individual observations each of which have some degree of credibility or confidence level. As these observations congregate within some processing environment, their value now becomes greater than the sum of individual observations. [7] The buildup of these observations into some cohesive picture is the discussion focus of this section. In early data communications, a data fusion type of concept was used in which voting was the means of verifying the quality or error-free performance of a certain transmitted message or its transmission medium. This entailed the reception of three status reports or three actual transmissions, and the examination of each to determine if they supported each other. The passing criterion was a two out of three agreement. This simple but straight forward method is still in use in many similar situations. The same method can be used to eliminate duplications, and to verify the existence of the same event over two or more sensors.

The next few subsections will discuss some of the techniques used to support data fusion tasking. These techniques range from the simple to the sophisticated, the costs of which are not necessarily proportional.

4.2.1 Classical Inference

Classical inference is the technique learned by every student of statistics. It has a broad range of application, but is limited to single events. Its use in data fusion where, by definition, system knowledge is determined by corroboration of multiple events, is limited to the relationship of that event

to the population mean of many events. The association or disassociation
of the event in question is referred to as the null hypothesis.

The null hypothesis tests the notion that there is no difference
between the observed event and the population to which the event is
attributed. [2] The rules for the test of the null hypothesis are indicated in
diagrammatic form in Figure 4-4. The Type 1 error occurs when we
reject a true statement. In other words, if we were to reject the assertion
that there are 7 days in the week, we would be making a Type 1 error. A
Type 2 error is known as the *false-positive* error. Here, the assertion is
false, but we are accepting it as true. For example, dice have come up
sevens five times in a row. We assert that the dice are crooked, but, in
fact, they are not.

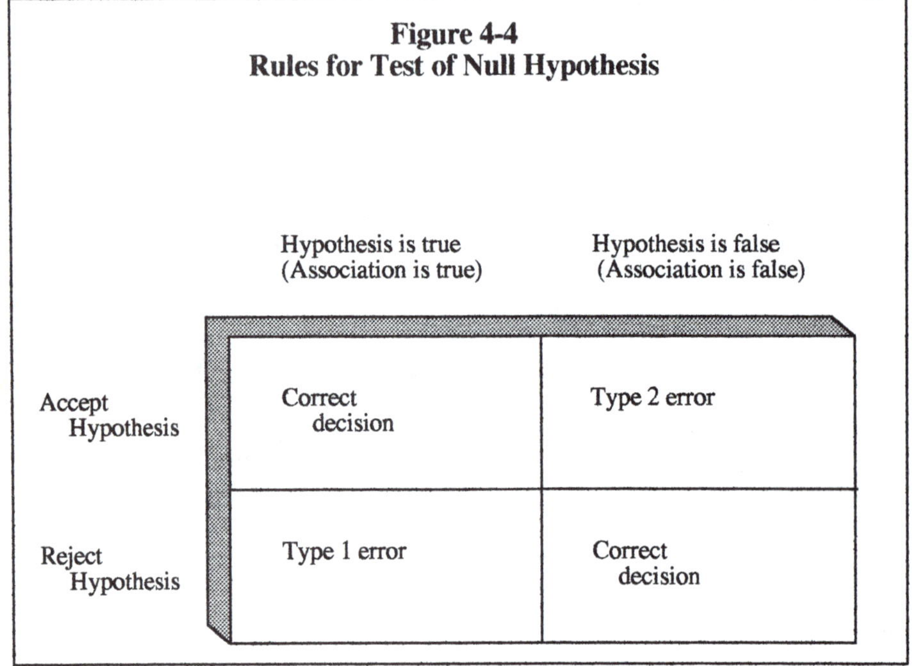

The rationale behind these tests is that there must be compelling
evidence that a difference exists between the observed event and the
population from which the event supposedly comes in order to reject the
proposition that a relationship exists. In order to test the null hypothesis
for any particular event arrival, the system must determine the mean,
standard deviation, and level of confidence at which a judgment can be
made. If, for instance, the mean is 1000, the standard deviation is 220, and

the confidence level is 2σ, then an incoming event with a value of 1445 is rejected as being part of the population just described for the following reasons.

$$z = x_i - m \Big/ \sigma$$

Where:

z = number of standard deviations observed event is from mean

x_i = value of observed event

m = ⟨ mean of the sample population

σ = standard deviation of the population sample

$$z = 1445 - 1000 \Big/ 220$$

$z = 2.02$

2.02 > 2σ confidence level; therefore, the null hypothesis is rejected.

Figure 4-5 illustrates the situation as it would apply to a 1σ test. In this case, the rejection of the null hypothesis takes place at a point above the 1σ limit. The same process can be used for those event values less than the sample mean. In such cases, a negative value for z indicates that the random variate is below the mean, and the evaluation makes allowances accordingly.

4.2.2 Bayesian Inference

Other, more sophisticated methods of dealing with data fusion problems entail the use of various types of identification features within each report, e.g., source, tagging, such as equipment type, probability, and confidence level indicators. Each of these is then compared to preceding or succeeding reports to determine which pertain to the same sensors. Those reports that fit certain categories are grouped together and analyzed. This analysis takes many forms, such as Bayesian inference or weighted averaging, to name a couple of examples.

Bayesian inference is an approach to data fusion based upon the recognition that the likelihood of certain events changes as more becomes known about the existence of certain other events. [2] The use of the Bayesian approach is implemented using the formulation provided below.

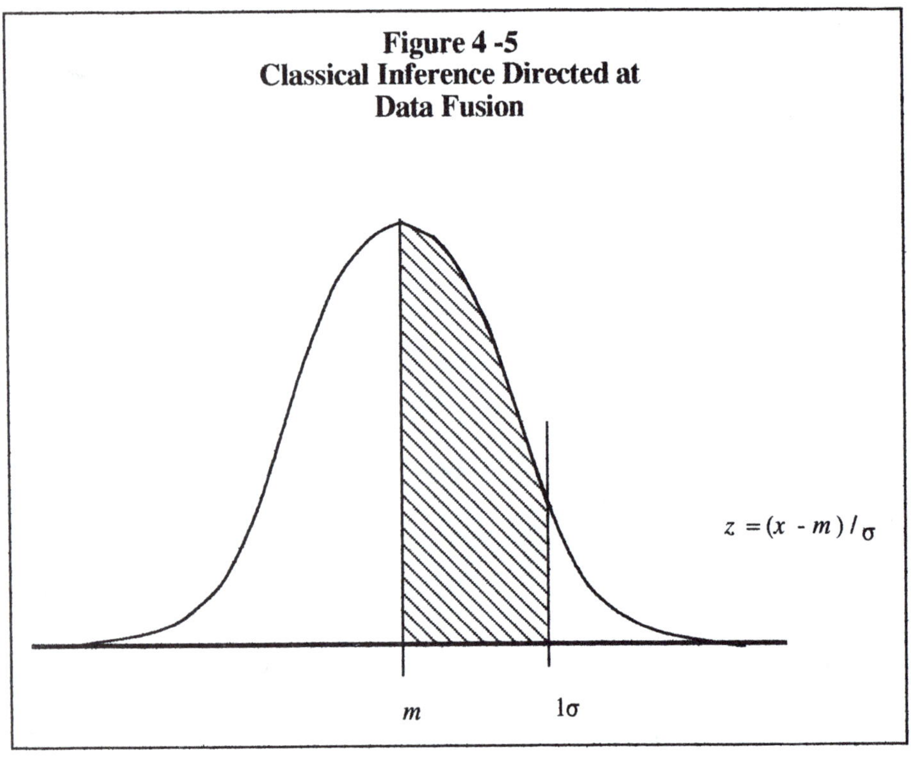

For 1 event and N hypotheses:

$$P\left(H_i / E\right) = P\left(E / H_i\right) P\left(H_i\right) \Big/ \sum \left[P\left(E / H_i\right) P\left(H_i\right) \right]$$

Where: E = the given event
                  H_i = any of n hypotheses from i to n

Thus, the probability that a particular hypothesis, or scenario, will occur given some single event occurrence, is a function of that event occurrence if the hypothesis or scenario really does materialize together with the probability of that event for each of all possible scenarios. For instance, if the event occurs, its likelihood of causation due to a specific scenario is a function of that scenario probability divided by all possible scenario probabilities.

4.2.3   Dempster-Shafer Method

The Dempster-Shafer method of data fusion was developed based on the independent work of the two individuals for whom the method is named. It represents a means to weigh the evidence for each cause that arises, a means to deal with the existence of uncertainty in the evidence, and results in a probability interval for the hypothesis developed. Additionally, the method has a broad range of application as do the previously discussed methods.

This approach to data fusion relies upon a relationship among evidence, belief, the distribution of that belief, and subsequent propositions about the probabilities in that domain. The Dempster-Shafer method is similar to classical inference in the sense that it defines an interval in which the null hypothesis, or the plausibility of a proposition, cannot be refuted. The situation is best explained by looking at Figure 4-6.

If, for instance, there is a piece of evidence that tends to support some proposition, and there is other evidence as well, then the two sets define an interval along the continuum of possibilities. The total probability space for all supporting and refuting evidence is 1. Further, if it can be shown that other evidence that supports the proposition results in a probability set of some value, then the sum of the two represents the range of plausibility, and the difference between the plausibility range and 1 now represents the probability of refutable evidence for any observation.

4.2.4   Fuzzy Sets

Fuzzy sets are a part of what is known as *set algebra* where set elements have a membership function, i.e., the membership in a set is defined as fuzzy and not boolean. [1] Fuzzy set inferencing is illustrated in Figure 4-7. An example of fuzzy set theory is as follows. Given that there are several types of packet data protocols being exercised in a system operation, and that these types are called classes of protocols. Additionally, there are parameters associated with each class called functions, with each of these protocols, such as packet sizes, header data, sequence numbers, etc. Assume that a packet report of certain observed parameters arrives. What are the class and functions of the observed packets?

The basic paradigm for fuzzy set theory is that observations arrive at some observation point. They are assessed, and determined to belong to one or more sets with some level of quantified value. The determination is made based upon an a priori identification (functions) of set membership (classes). Also, the set algebra, or operator methods, must be defined.

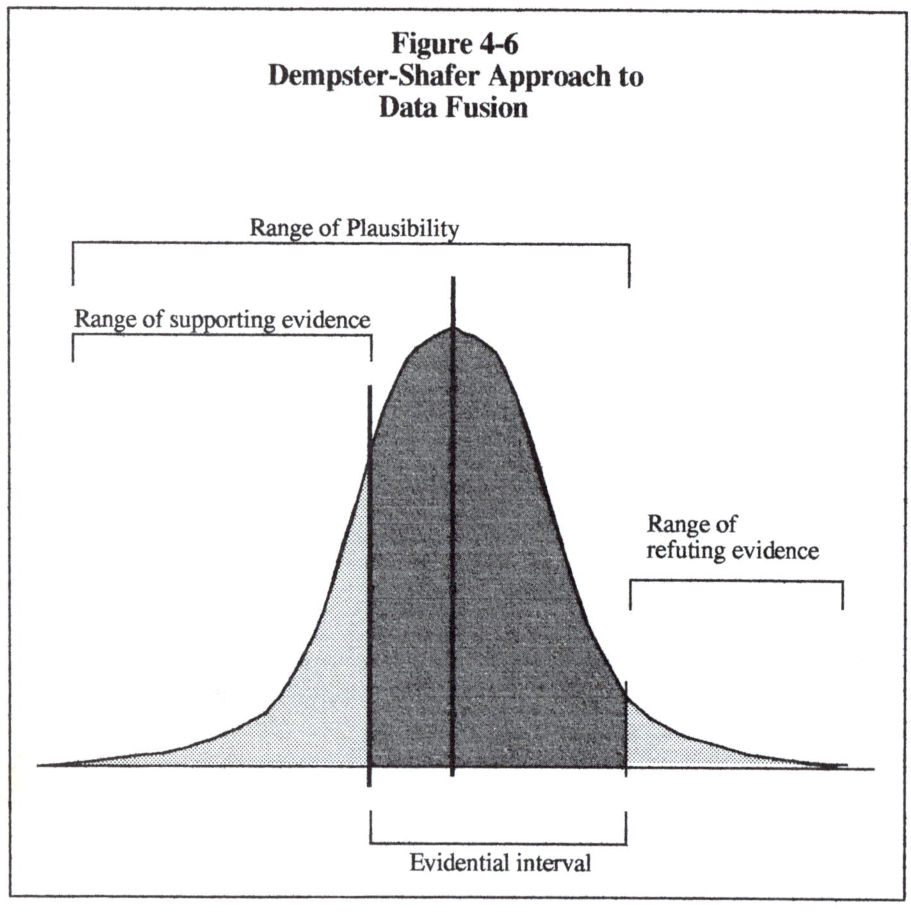

The underlying premises of fuzzy set theory are as follows. First, human understanding is not binary, black or white, etc. There are degrees of gray area, or ambiguity. Second, human expression is not limited to well defined terms, such as *one point zero*. Often, the concept is defined as *roughly eight, heavy, about here*, etc. Such specifications are not defined with the usual math precision, but are part of the normal human specification of the situation. These expressions and others like them are part of what is known as *linguistic variables*. Such variables have no absolute range of inherent variation. Their value is solely dependent upon the interpretation of the user of that variable.

For instance, what is high for one person, in absolute measurable terms, may not be high to the next person questioned. Third, the imprecision of a notion or statement is an intrinsic property of language, and can be further characterized as: (a) not an approximation to the truth, but rather simply an honest evaluation; (b) not a failure to comprehend, but rather as close an understanding as possible; (c) an admission that some notions about the world are basically imprecise.

Figure 4-7
Fuzzy Set Inferencing

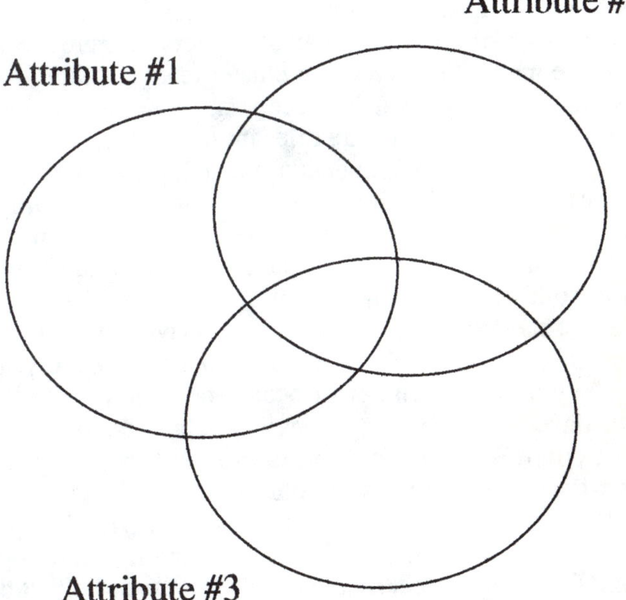

Set A = (Attribute #1, Attribute #2)
Set B = (Attribute #1, Attribute #3)

Set A and Set B are not identical, but they have some attributes that can be equated in order to show association.

Fuzzy sets are sets of ordered pairs, e.g., classes and functions. They are also part of a membership, and the membership is also a function of the parameters of the ordered set. An element, or one of the ordered pair, may be partly in one set and partly in another. Thus, one of the classes of data may include packet data as opposed to a data stream. So, class information may relate to both packet and continuous data transmission. The other element of the ordered set may be the function which, in this case, may be continuous. Therefore, the fuzzy set would include a continuous data stream that might be indicative of a fax transmission, for example.

4.2.5   Cluster Analysis

This technique involves the assembly of samples into natural groupings. The inputs are a group of n-dimensional observations, and the outputs of the process are a set of groupings, called clusters. Figure 4-8 illustrates the end result of the method involved. In cluster analysis, there are a couple of critical assumptions that must be accepted. One assumption is that the components of the observations are relevant to the question being asked.

These observation components are usually the comparison attributes against which the other cluster observations are being measured. Relevance means that the attributes can be directly or indirectly implicated in the answer of the question being asked. An example of such a question might be the collection and analysis of observations that are made during the operation of a LAN. The question to be answered may be whether any or most of the errors are related to disturbances in the public network supporting the operation. The error occurrences may be analyzed by type and time-of-day. The error type may give indications of whether the network is a contributing factor, and the time-of-day could help correlate error types with network congestion during a typical time frame.

The other assumption is that the user is correctly interpreting the results of the analysis made. This is probably not always a good assumption. Thus, the correct association of cause and effect depends upon assembling the right attributes that relate to the observations, along with the relationships between the clusters and the effects. In other words, the clustering of observations is dependent upon the attributes chosen, and the interpretation of results assumes that there is a correlation between the attributes and the available causes.

Cluster analysis may use many and varied techniques, such as sorting algorithms that separate observations into their constituent groups. One of the advantages of cluster analysis is that it does not necessarily require statistical information about the data or events, because the approach is ad hoc in nature. The sorting algorithms are also useful in developing relationships between groups of variables without the need for a priori models. There are two basic types of sorting algorithms used for cluster analysis, these being hierarchical and non-hierarchical. Indicated below are examples of these two methods.

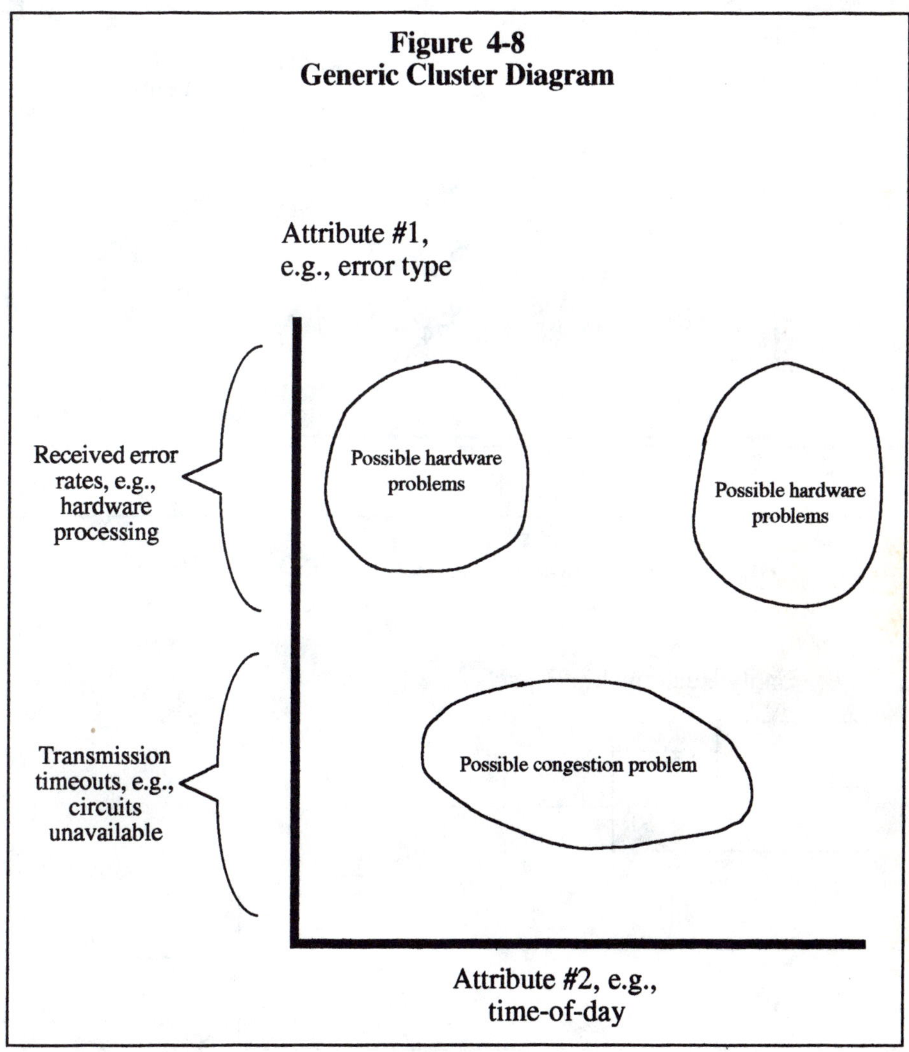

Figure 4-8
Generic Cluster Diagram

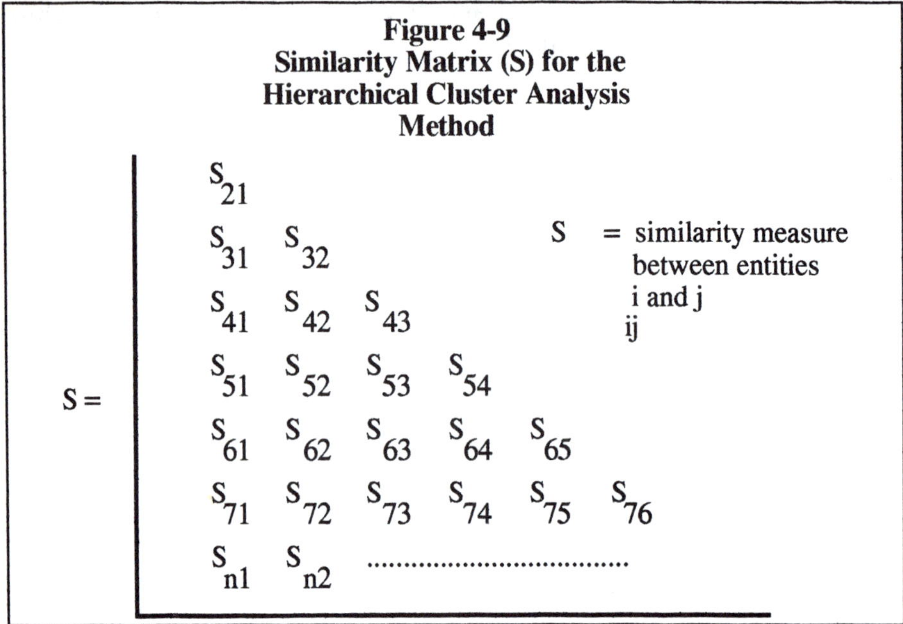

$$S = \begin{array}{cccccc} S_{21} \\ S_{31} & S_{32} \\ S_{41} & S_{42} & S_{43} \\ S_{51} & S_{52} & S_{53} & S_{54} \\ S_{61} & S_{62} & S_{63} & S_{64} & S_{65} \\ S_{71} & S_{72} & S_{73} & S_{74} & S_{75} & S_{76} \\ S_{n1} & S_{n2} & \cdots \cdots \cdots \cdots \cdots \end{array}$$

S_{ij} = similarity measure between entities i and j

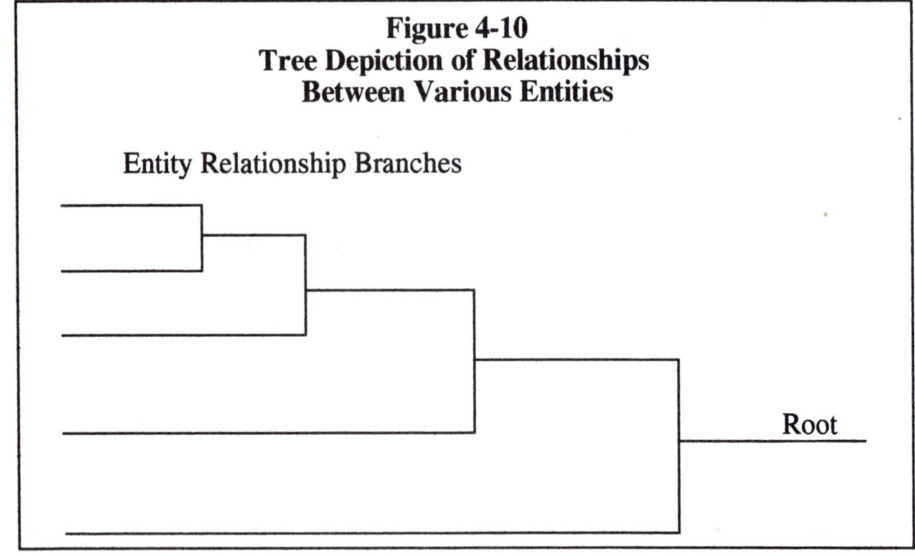

Hierarchical Clustering Method

One can consider the existence of a similarity matrix that describes the strengths of all paired relationships among entities being investigated, as either variables or data sets. Hierarchcial cluster methods operate upon the similarity matrix to construct a tree depicting specified relationships among these entities, as shown in Figure 4-9. These relationships are the attributes that can be analyzed to derive the root identification of the entity in question. Figure 4-10 depicts a set of representative relationships between branches to illustrate the dependencies between different similarity matrix entries. The tree branches indicate relationships that can be assembled in the matrix.

Such methods operate using a basic kernel, such as the definition of the Euclidean distance, as the similarity measurement parameter. This particular measure is defined as follows.

$$\text{Euclidean distance} = D_{ij} = \left\{ {}^n\Sigma_{k=1} \, (x_{ik} - x_{jk})^2 \right\}^{1/2}$$

Where: $n =$ number of measurements
           $x =$ value of data vector at measurement k

The overall strategy for the operation of a generalized sort process is as follows:

(1)   Begin with intersecting clusters, each consisting of one entity.

(2)   Search the similarity matrix for the most similar pairs of clusters; e.g., the similarity criteria are defined as S_{pq} using the p and q search parameters.

(3)   Reduce the number of clusters by 1 through merging p and q. Reduce the similarity matrix to order n-1.

(4)   Repeat steps 2 and 3 (n -1) times.

Of the hierarchical methods, there are three major approaches, these being linkage, centroid, and the error sum of squares (variance method).

(1) Linkage Approach
The linkage approach calls for an initial linkage of elements according to some maximum linkage criterion, usually a shortest distance parameter. Succeeding links are made using this criterion until no additional linkages can be made. The resulting set of clusters defines the clustergram based upon the criterion used.

(2) Centroid Approach
Using this approach, the analyst attacks the scattergram by defining centroids. The centroid establishes a minimum weighted vector to which each member of the group can be assigned.

(3) Variance Approach
In the variance approach clusters are merged at successive stages in order to maximize an objective function that reflects an a priori purpose by the user. In other words the merging of clusters results in a minimum increase in the group variance.

Nonhierarchical Method

In the nonhierarchical method the process of action is a little simpler. The analyst controls the partitioning of data units into clusters, and the altering of cluster memberships, in order to obtain a better partition. There are also three types of nonhierarchical clustering methods.

(1) Arbitrary Partitioning
The analyst selects an initial partition of data into n clusters. As the sample sizes are allocated, or the initial cluster constellation is built, the analyst adjusts the clustering parameters to suit a "better fit." Obviously, the fit and its suitability are qualitative measures that can only be judged by the analyst. The influencing factors are the density and membership sizes of the clusters.

(2) Analytical Strategy
A pseudo-analytical approach is one in which numerical analysis methods are used to justify or reduce the observations to groupings. An example of this method is as follows:

(1) Choose a set of seed data points to be located within the scattergram.

(2)   Allocate each and every observation to the nearest seed data point within the total array of observations.

(3)   Now compute new seed data points as a function of the centroids of the clusters established during step 2.

(4)   Iterate through steps 2 and 3 until desired or acceptable convergence is is obtained.

(3) Other Methods

Other methods include several nearest centroid approaches that rely upon a fixed and given number of centroids that in turn will generate a fixed number of clusters. Another method involves a number of randomly spaced centroids that pass or fail as centroids based upon whether observations cluster around them as a result of the iteration process described above.

4.2.6   Estimation Methods

Estimation theories yield more information about the parameters than the stories about the entities being affected. The kernel processes behind the estimation theory of data fusion and its usage include the least squares fits, weighting, and figure-of-merit. Thus, system observations are collected, and state vectors are generated to yield a picture of the event(s) affected.

Point Estimates

A point estimate is a statement of reality that may or may not be substantiated by the facts. People tend to deal with point estimates easily, because they represent less work to remember or define. Unfortunately, point estimates are misleading unless accompanied by qualifiers, such as their standard deviation, level of confidence, sample size, etc. Point estimates should be avoided unless accompanied by the relevant qualifiers.

Weighted Estimates

Estimation theory is a method of assigning weight to a sample that is supposedly taken from a population of similar characteristics. [2] It can be expressed as a function giving weight to both sample and population, as follows.

Estimate $= \omega \cdot x + (1 - \omega) \cdot \mu$

where: ω = weight given to sample
              x = sample mean
              μ = population mean

The weight can be either a percentage weight or a relative importance weight. In either event, it must fall in the range of 0 to 1. For the condition that the square of the error of the estimate be as small as possible, it can be shown that the weighting function is as shown below.

$$\omega = \frac{n}{n + \sigma^2 / \sigma^2_\mu}$$

where: ω =   weighting factor assigned to the sample
              n =   number of samples upon which the weight is based
              σ^2 =   squared variance of the sample
              σ^2_μ =   squared variance of the population

The implications of this estimation process are important. If there are no samples with which to evaluate the weighting function, the weight is zero, and all estimates will be based upon collateral information; i.e., the estimate will equal μ, the mean of collateral populations. As n increases, the weight is shifted to the sample, but slightly, because the sample variance is large compared to the population variance. As n continues to grow, the sample variance begins to approach the population variance, and the ratio defining the weight approaches 1. At this point, the bulk of the weighting function is shifted to the sample mean and away from the population mean.

Chi-Squared Estimation

Many times the estimation of the actual population mean is difficult to bracket. There have been several methods devised to overcome this short coming, probably the most popular of which is the *chi-square* distribution. This method of sample variance estimation is implemented as follows. An inequality is set up to modify the sample variance based upon the chi-square distribution. The inequality is a comparison of two ratios that utilize the sample size, sometimes called the degrees of freedom, the sample mean, and the Chi-Square distribution values. These are arrived at by first defining the confidence limit required, and then accessing the chi-square distribution table to determine the areas corresponding to those required confidence limits. The process is solved as shown below.

$$\left[d.f. \cdot s^2 / X^2 \right]^{1/2} < \sigma < \left[d.f. \cdot s^2 / X^2 \right]^{1/2}$$

where: $d.f.$ =   degrees of freedom, i.e., sample size -1
         s =   sample standard deviation
         X^2 =   chi-square distribution function
         σ =   estimated population variance

Sample Size Estimation

Another estimate revolves around the sample sizes required to predict the confidence of results to be obtained. In other words, we may want to know what sample size, given an acceptable confidence error of say 3 percent (0.03), we need to collect in order to be certain to some level of confidence, say 95 percent, that the true population mean will not deviate from the sample mean by more than the error tolerance specified. This figure can be arrived at as follows.

$$n = p(1-p)\left[z_{\alpha/2} / E \right]^2$$

where: n =   sample size required

         p =   point probability of the event

         $z_{\alpha/2}$ =   confidence limit imposed

         E =   acceptable error allowed

Thus, if the probability is guessed to be 1/2, the confidence limit to be 95 percent, or 1.96 for a normal distribution, and the error allowed to be 3 percent, then the estimated sample size required is:

$$n \;=\; 1/2(1/2)\Big[\,1.96\,/\,0.03\,\Big]^2$$

$$n \;=\; (0.25)\,\Big[\,65.33\,\Big]^2$$

$$n \;=\; (0.25)(4268)$$

$$n \;=\; 1067$$

Population Mean Estimation

Probably the best known estimate technique is that used for the estimation of the mean of a population, given the sample mean already obtained. The population mean estimation method was developed by W.S. Gosset, who used the pen name *student*. The technique incorporates the sample mean coupled with the confidence level to be imposed, the sample standard deviation, and the sample size. The *student-T distribution* is well known for its utility as an estimation tool, and is used in the following formulation context.

$$x - z_{\alpha/2} \cdot \frac{s}{(n)^{1/2}} \;<\; \mu \;<\; x + z_{\alpha/2} \cdot \frac{s}{(n)^{1/2}}$$

Where:

$$
\begin{aligned}
x \;&=\; \text{sample mean} \\
z_{\alpha/2} \;&=\; \text{confidence level; e.g., 95\% is 1.96, 98\% is 2.33,} \\
&\qquad \text{and 99\% is 2.58} \\
s \;&=\; \text{standard deviation of the sample} \\
n \;&=\; \text{sample size} \\
\mu \;&=\; \text{population mean}
\end{aligned}
$$

Thus, a problem where the question is that of determining the population mean based upon the sample mean where the sample size, standard deviation, and confidence level are known, can be solved easily. The confidence level stipulated dictates a range of accuracy in the prediction of the population mean solved.

Estimate of Proportions

Many times, in evaluating network performance, estimates of proportions are necessary in order to make the proper judgments. The Bayesian approach is used in those cases where sufficient direct evidence of the true statistics is unavailable. The basic formulations for this approach to estimation of proportions are:

$$\text{Estimate} \; = \; \omega \cdot \chi \big|_n + (1 - \omega) \cdot p$$

$$w \; = \; \frac{n}{n + \left[p(1 - p) \big/ \sigma^2_p \right] - 1}$$

where:

ω = weighting function for direct observations

χ = number of successes or errors suffered during the sampling period

n = sample size

p = probability of the collateral information or target statistic

σ^2_p = standard deviation of the collateral information or target statistic

By way of example, we might consider a performance situation in which a target error rate of 1% is desired, and we can call this p. The number of observed errors, χ, seen during the sampling period is 6, and the sampling size observed is 100. The standard deviation, σ_p, is determined to be 0.005. The weight, ω, works out to be 0.005. In working through the formulas given, the weight reduces to 0.20, or 100/495, and the estimate of the actual probability is a combination of the observed data and collateral data or, in this case, the target probability, which is $p = 0.01$. The estimate is $(0.20) \cdot 6/100 + (0.80) \cdot 0.01$. The result is 0.020 or a 2% probability of getting an error in actuality as opposed to an assumed error of 1%.

Summary:

Data fusion is key to the organization of and confidence level attached to the data presented to an analysis or decision support system. It is very closely associated with both the decision support and the sensor processing functions. Sensor data can only become meaningful after it is sifted, sorted, and classified for use further up the decision making command. The key to effective data fusion is the association of like reports. This is the data integration function mentioned earlier. The key features of the data fusion process can have a very positive effect upon the simplicity, quality and operating costs of the target system.

References

(1)　Ericson, C., L. T. Ericson and D. Minoli (eds.), *Expert Systems Applications in Integrated Network Management*, Artech House, Inc., Norwood, MA, 1989.

(2)　Freund, J. E., *Modern Elementary Statistics*, Prentice-Hall, Englewood Cliffs, NJ, 1967.

(3)　Hall, David, and James Llinas, *Data Fusion*, Technology Training Corporation, Los Angeles, CA, 1986.

(4)　Harris, C. J., and I. White (eds.), *Advances in Command, Control & Communication Systems*, Peter Peregrinus Ltd., London, UK, 1987.

(5)　Hopple, G.W., *The State of the Art in Decision Support Systems*, QED Information Sciences, Inc., Wellesley, MA, 1988.

(6)　Miller, I., and J. E. Freund, *Probability and Statistics for Engineers*, Prentice-Hall, Englewood Cliffs, NJ, 1965.

(7)　Singleton, W. T., "Man-Machine Aspects of Command and Control," in *Advances in Command, Control and Communication Systems*, Harris, C. J., and I. White (eds.), Peter Peregrinus Ltd., London, UK, 1987.

(8)　White, F. E., and J. Llinas, "Data Fusion: the process of C^3I. (command, control, communications and intelligence)," *Defense Electronics*, vol 22, p 77, June 1990.

(9)　Wilson, G. B., "Some Aspects of Data Fusion," in *Advances in Command, Control and Communication Systems*, Harris, C. J., and I. White, (eds.), Peter Peregrinus Ltd., London, England, 1987.

Chapter

5

Decision Support Systems

Chapter Highlights:

Decision support systems (DSS) are derived from command and control systems previously developed by the military over the past 50 years. They embody the same features in so far as the data analysis is concerned and to some extent, include many of the same types of decision-making algorithms and data collection processing found in military systems. This chapter addresses these systems in terms of their architectures, assessment capabilities, feature design, and competitive advantage issues. DSS is yet another step in the evolution of aided decision making that started in the 1950s with data manipulation, known as electronic data processing (EDP), then evolved to data analysis, known as management information systems (MIS), and finally to the present form of multi-source, knowledge-based evaluations known as decision support. Such systems tend to be somewhat unstructured in both user interface requirements and analysis logic.

5.1 Perspectives

The previous chapters have discussed the front-end features associated with smart network management systems, namely sensors and data fusion. [14] The features that turn this collection and integration of data into meaningful and useful information are referred to in the aerospace-defense sector as decision support systems (DSSs), or executive information systems (EISs), as they are known in civilian-commercial environments. [13] These two terms and abbreviations will be used interchangeably in this chapter, however. DSS/EIS is viewed in some contexts as a simulation dealing with "what if" questions. Also, EIS may be construed as a means of tracking resource usage, while DSS may be interpreted as a means of requesting direction. By varying the input parameter values, or the model, the consequences can be determined, and a family of answers can be developed. Multi-variate changes can also be evaluated in many cases using the simulation approach.

Whether or not DSS/EIS is classified as a simulation or an on-line control, it functions to provide the user some degree of oversight with regard to the telecommunications system, and in that regard, it is a valuable network management tool. As such a tool, the DSS/EIS can be used in the background to play out scenarios about the functioning of the network, and the various faults that may befall the system.

The idea behind the DSS is to provide the observer with the best view of the situation, resources, actions, and options available to him or her that can be constructed. In this regard, DSSs are not a substitute or alternative for network management facilities, but rather they support the specific five functions defined by the OSI model for the total network management problem. These functions are: (1) fault monitoring; (2) performance monitoring; (3) configuration management; (4) accounting management; and, (5) security management. Each of these will be discussed in more detail in later chapters.

Decision support, however, targets the three main issues within the realm of network management, these being planning, performance, and maintenance. [10] Shown in Figure 5-1 is the relationship between the planning, performance, and maintenance aspects of the network management problem and the basic operation. Fault and operational data analysis feed the fault evaluation module. This information is used to update the system configuration module. The system configuration status and the operational data analysis modules feed performance evaluation that in turn provides input to the control implementation module that provides instructions to the system for reconfiguration. Meanwhile, the periodic updating of the system configuration module activates the planning evaluation module, which also contains the system goals. The network configuration files contain connection routing tables, equipment inventory, network topology, and element capabilities.

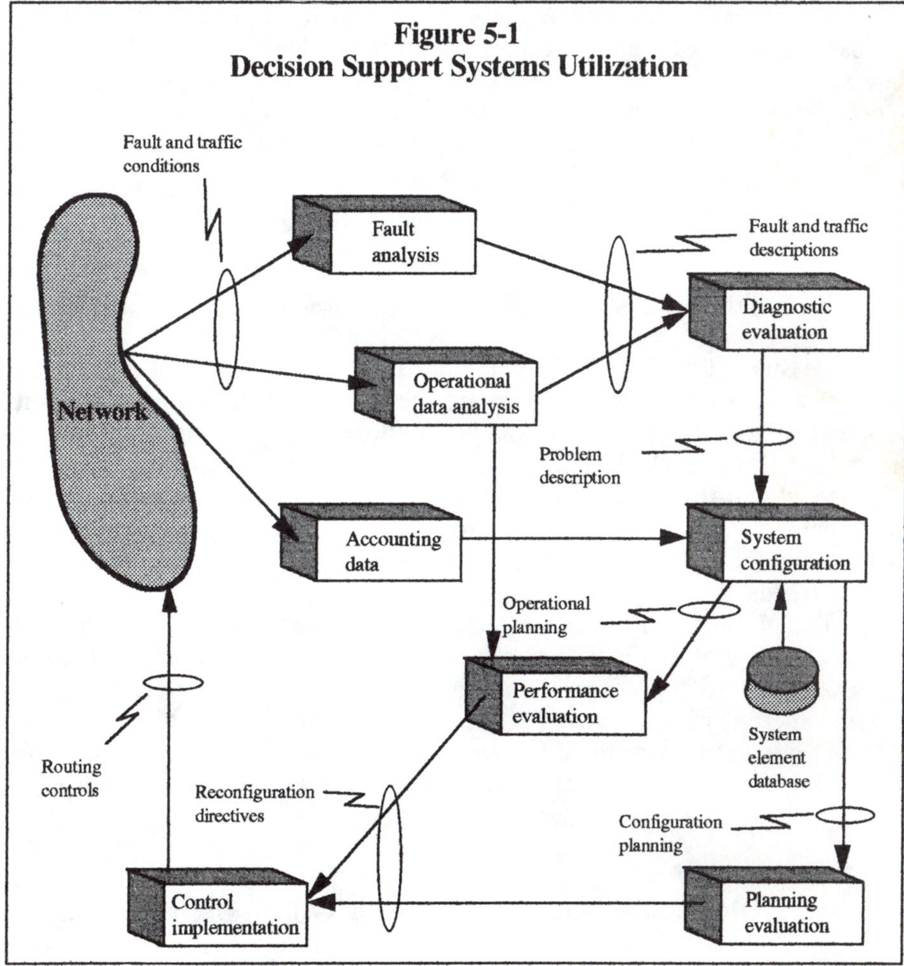

DSS/EIS is not intended to be an entirely simulation-style tool, as will be seen at the end of this chapter. Often, in the early stages of DSS/EIS development, the capability is concerned with tying together remote locations via communications links. Additionally, the objective is to provide users with the accesses to multiple databases. And lastly, the system is completed by supplying report generators. Report generators may undergo a lengthy or even indefinite development period where the generator progresses from data presentation, to data transformation into parameters of value, to simulations, and finally to intelligent applications. Such applications may use rules, heuristics, or any of a number of references to generate suggestions, or decision-action situations.

In this chapter we look at the use of DSS/EIS to perform information assessment and plan generation, evaluation, and execution. The tools required to perform these tasks include the system architecture, the user interfaces needed, and the various analysis aids. All of these tasks involved and the tools needed will be addressed here. The next few sections will discuss briefly some of the distinctions associated with decision or executive information support.

5.1.1   Time Basis for DSSs

The assessment of, and recommendations for present and future actions associated with a particular subject are the domain of the decision support or EIS system. Present action analysis and decision making involves the assimilation of current data received into the system. Such data is can be received in real-time or via some other slower methods, such as telephone relay, hardcopy, manual collection, etc. Future action analysis and decision making involves the use of simulation tools to reach conclusions based upon current trends, and projected outcomes.

The buildup of DSS/EIS systems in a decision-oriented environment requires the use of modular concepts. The use of building blocks is usually necessary due to the complexity of the processes involved and also to provide incremental features for the user. Finally, it is critical that the DSS provide the right vehicle for decisions. It is a given here that decision support will not replace decision makers in human form, but it can be a powerful assistant to the human decision maker.

5.1.2   Distinctions Between DSS and Other Decision Aids

MIS, as a tool set, is one of the closest such tools to DSS, and therefore a source of confusion for someone involved in this area. MIS appears to be more concerned with the efficiency than the effectiveness of the measures derived. MIS takes reported data and distills it down to needed measures of operation. Additionally, MIS appears to be more occupied with current and past data sets than developing future or predictive models.

Another type of decision aid is simulation. Simulation is a generic term applied to any procedure that utilizes probabilities, obtained from current or past performances, that are applied to projections of future performance. Simulation has reached significant heights of sophistication with graphic representations, interfaces to other programs, and artificial intelligence features. Simulation tools are very sophisticated these days, and provide some of the valuable features of DSS capabilities.

5.1.3 The Decision Support Cycle

The decision process is the culmination of several steps aimed at affecting an outcome of some objective. This generalized process is illustrated in Figure 5-2. The utility of developing an effective DSS lies in the results it creates, i.e., its validity, the time it takes to generate an answer, and its consistency, i.e., reliability of its results. If the results are within an acceptable range of error, the system is considered to meet a certain set of expectations, sometimes known as specifications. If the system turns out the same set of results time-after-time, it is demonstrating an agreement with functional requirements. And, if it meets response time constraints, it is meeting both specifications and functional requirements. The first part of the process involves data observation (sensors), and data collection (fusion) of events within the environment, two subjects addressed in earlier chapters.

The steps following the decision action are part of the executive function, i.e., the translation of decision into a command structure, communication of orders to the command, the execution of orders, and the responses elicited by the execution of orders upon the environment. The communications infrastructure is usually the object of initial interest. In real-life implementations, such as those cited at the end of this book, many such DSSs started with ad hoc communications systems implemented to satisfy increased communications requirements between functional and/or geographical units. At some point later, DSS/EIS capabilities were conceived and developed separately.

There are a few points to be made relative to this figure. First, the decision support system must be able to receive data relevant to the decision environment. The selection and use of the right input data sets, is critical to the effective generation of decision alternatives. Second, the key to the success and evaluation of the DSS is its ability to measure the impact of previous decisions from responses obtained from the environment. Third, some form of communications medium is needed to put the decisions into action. No DSS worthy of the name can be effective without a collection facility, feedback capability, and a means to put the decision into effect.

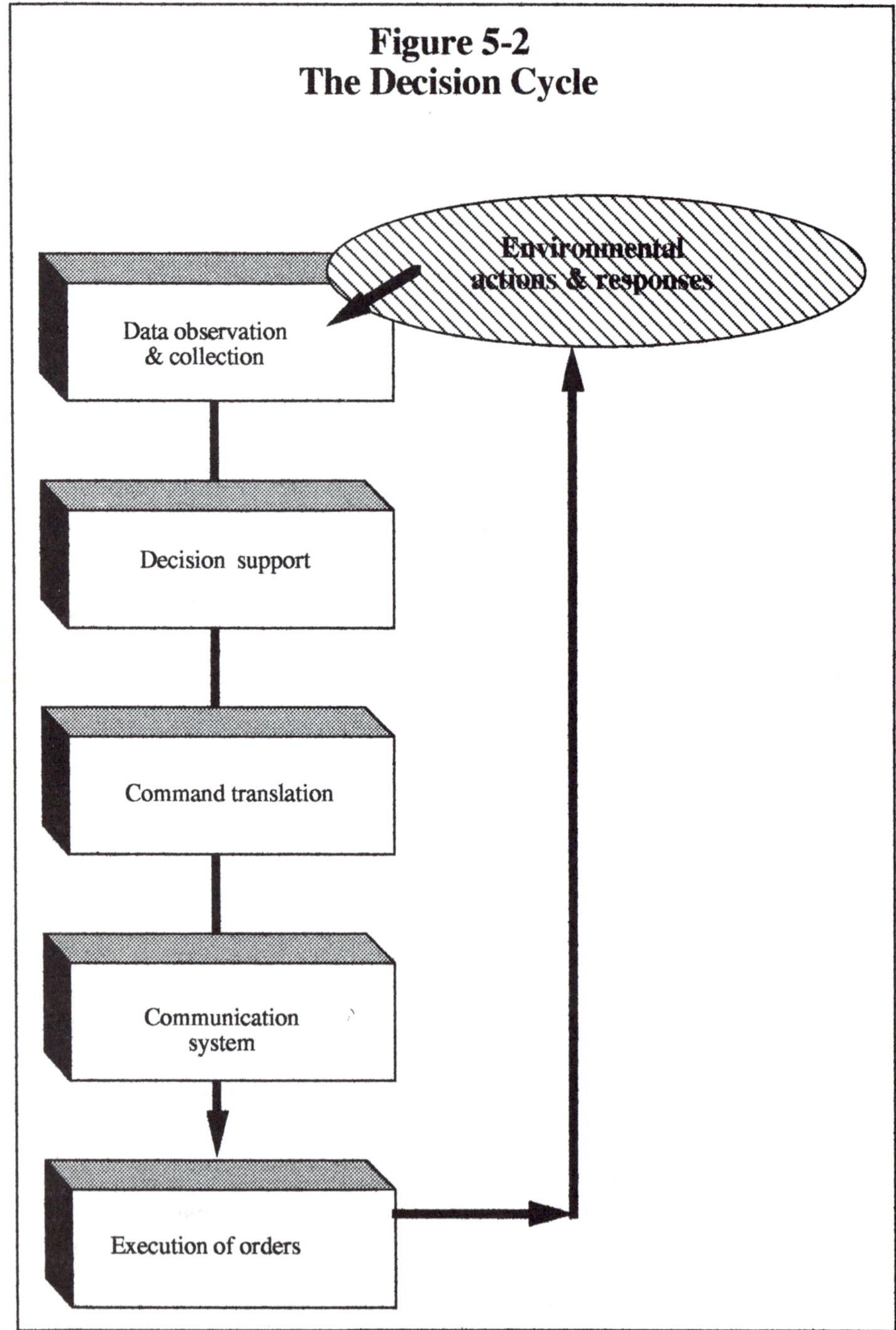
Figure 5-2
The Decision Cycle
Environmental
actions & responses
Data observation
& collection
Decision support
Command translation
Communication
system
Execution of orders

5.2 Architectures

5.2.1 Generic Requirements

The challenge for DSSs is not that they must keep pace with the increasing influx of data, but rather that they must do a better job of making sense of the data that is collected. The typical DSS will have a series of attributes associated with it, some of which are:

(1) timeliness
(2) filtering
(3) user aids
(4) interfaces to other systems
(5) intelligence (flexibility, inference, pattern recognition, judgment, fuzzy logic, etc.)

The degree to which values and ranges are assigned to these attributes, will determine their utility in the decision part of DSS. The variety of attributes available for analysis will dictate the support part of the DSS, and the breadth of sensors providing input will give meaning to the system part of the DSS expression.

Some brief definition of these attributes is provided below.

(1) Timeliness

Timeliness, as an attribute, means that the incoming data must bear some resemblance to other incoming data sets in order for proper evaluations to be made. The timing of inputs often provides valuable clues as to the relationships between reports, the validity of reports, or their reliability.

(2) Filtering

Filtering is the ability of the DSS to make selective judgments about information, and/or minor decisions, the object being to separate unnecessary information from what is crucial to the decision process. Many times, the distinction between necessary and unnecessary information helps draw lines about the credibility of the process or the results obtained.

For instance, the value of an operation may be its safety impact as opposed to cost control or revenue impact. The latter measures may, therefore, be somewhat unnecessary or irrelevant when evaluating the basic value of the system. Reports about such issues may, therefore, be filtered out so that the system can evaluate reports of interest to the user.

(3)   User Aids

User aids are the tools by which the user can distill and clarify information into something meaningful, and therefore, something of importance to the mission of the DSS. User aids give the system a certain utility so that the intended user can exercise the system effectively. These aids include the presentation formats, user interfaces, the analysis tools, correlation capabilities with other data, and interfaces to other systems.

(4)   Interfaces

Interfaces to other systems are the connectivity issues, both physical and functional, that allow data and decisions to be conveyed to other systems, and that make reception of data possible into the system. The ability of the system to interface to other systems also gives it additional power to draw upon resources not part of its basic makeup. This interfacing, again, requires communication means such that the sharing of data lends increased power to each unit, redundancy within the network, and access to databases not possible with isolated units.

(5)   Intelligence

Intelligence is the internal composition of the DSS that allows it to exercise some degree of problem solving, and range of action. Intelligence is usually equated to such features as artificial intelligence, machine vision, and robotics.

5.2.2   Types of DSSs

The various types of decision support systems are described and delineated here. One such type, the MIS system, was mentioned above. It was one of the first organized decision structures to be developed in the early era of computerized data manipulation. Since then, more sophisticated means of decision support have been developed, but MIS systems still exist in most companies as decision aids.

(1)   Report generator systems (similar to MIS)
(2)   Data analysis systems
(3)   Analysis information systems
(4)   Accounting models
(5)   Representational models

(6)   Optimization models
(7)   Suggestion models

(1) Report Generator Systems
The first and primary information need is access to data items. MIS was developed to meet this need. Implicit in this delivery is the immediacy of the data item availability. Thus, MIS stresses on-line accesses to the supporting databases, with rather basic user interfaces to bridge the databases and the report generation.

(2) Data Analysis Systems
Such systems are developed to perform very specific analyses. Users could manipulate the data to generate very specific reports. The range of analyses is limited to the applications software of the system.

(3) Analysis Information Systems
This type of system utilizes relational databases and modeling capabilities to create the data measures required. Such systems have some latitude as to the nature and form of measures created.

(4) Accounting Models
Accounting models use definitional procedures to calculate the outcomes of particular circumstances. These systems are used to calculate the consequences of possible financial decisions.

(5) Representational Models
Modeling used in this type of system is directed at attributes and items that are not necessarily accounting-oriented in nature. In such systems accounting terms are derived as opposed to being entered as parameters, as would be the case with accounting packages.

(6) Optimization Models
These systems provide the ability to profile situations mathematically that provide the best answers among the modelled circumstances. Examples of such capabilities are the required price and inventory points necessary to achieve customer demands.

(7) Suggestion Models
Suggestion models offer recommended courses of action as opposed to offering the tradeoffs that are part of optimization models. This is an outcome approach as opposed to an evaluation-based approach to modelling.

5.2.3   Design

The basic configuration for any DSS/EIS includes the following elements:

(1)   Interface to the Outside World

This feature can involve environmental or process sampling, database interfaces, and interfaces to other computational systems. External interfaces for network management systems usually involves an array of sensory inputs provided via communications services, e.g., private line, packet network, and dial-up facilities.

(2)   Buffering and Storage

The data stream from the external interfaces must usually have a temporary repository while it is waiting for processing. Such temporary storage is provided by the buffer/storage control and memory area of the computer. At this point, some pre-processing is also performed on the incoming data sets to eliminate such problems as duplicate and unreliable data.

(3)   Algorithmic Manipulation

The main DSS/EIS activity is concerned with the procedural control and transformation of the external data stream. This process may range from summary data reporting, solutions of equations and subsequent reporting, multiple solutions to equations used to generate graphical representations with the help of graphics generators, simulation systems to allow projections and to play what-if games, and rule-based derivations to isolate likely solutions.

(4)   Human-Machine Interface

This element has already been discussed to a large extent. In summary, the human action is directed at assuring that the required input data is available, the appropriate processing is performed, the parameters and answers obtained are reasonable, and that the proper output or results are reported. Such a set of activities can either be labor intensive, or with the help of expert systems technology, human judgment can sometimes be replicated and substituted effectively.

These elements are organized and illustrated in Figure 5-3.

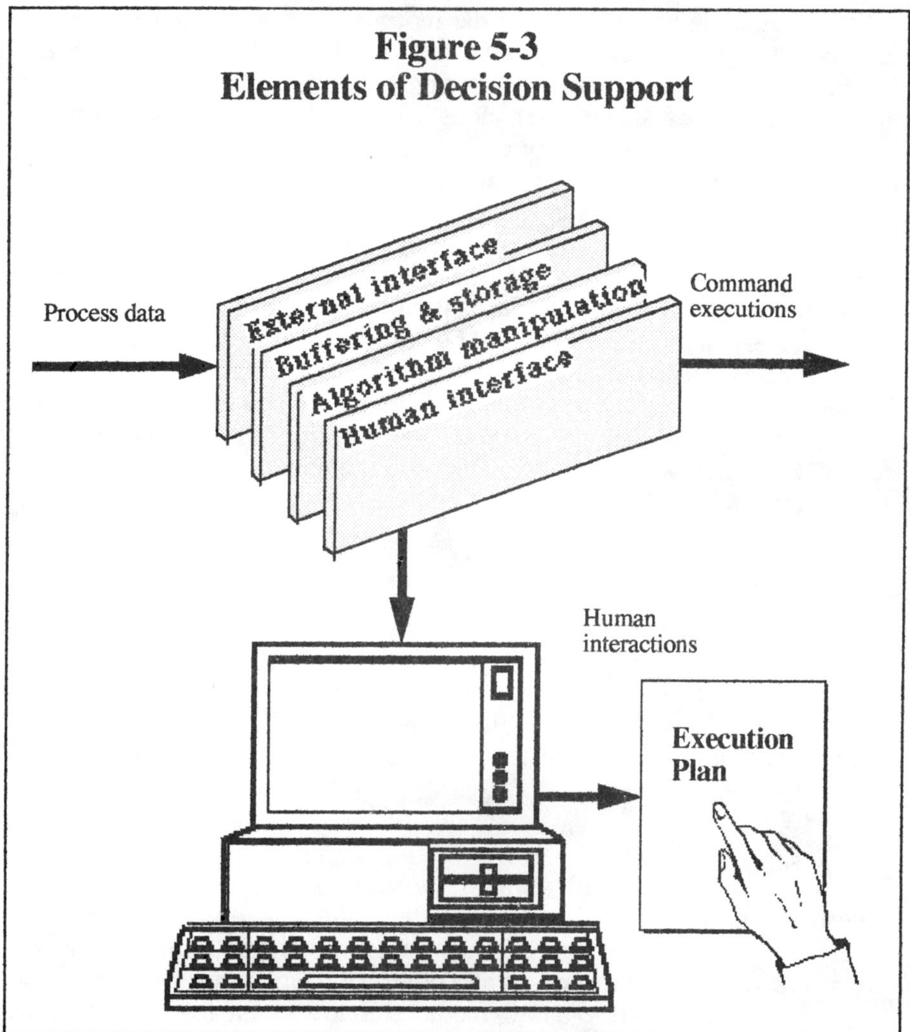

5.2.4 Generalized Architectures

The general architecture of the typical DSS is illustrated in Figure 5-4. [8] The external environment supplies the raw data needed for analysis. Trends and historical arrays of these data are stored in the database and available for manipulation through the use of the database management system. The model base provides a replication of reality as seen from the user's point of view. The model is a set of rules (see The Expert System below) that expresses how the system should function, and how its operation is affected when it doesn't function properly.

External stimuli arriving at the input to the system are examined for correlation with the current status of the model as driven by the data items in the database. The decision support interface assembles the most appropriate view of the situation along with options to design or redesign a solution. To determine exactly what is appropriate is the challenge for the decision support interface. Presumably, there is a need to provide enough information in a short period of time and to package it in as meaningful a form as possible for the user to digest and react to.

It is interesting to note that the technology is and has been available for many years to supply sophisticated decision support systems. Unfortunately, the pressure has not been there to force the development of these tools until recent years partly due to the lack of competitive influences or lack of complete understanding as to how these tools relate to a company's success. Now, however, that the stimulus is present, there are several problems with which the decision support designer must grapple. These problems are briefly outlined below:

(1)　Technology
The technology systems needed or desired to solve the network management problem are many and varied. The technology problem usually revolves around the questions of speed, capacity, reliability, physical and functional connectivity, and access to data.

(2)　Behavior
Decision and general psychological theory are areas of investigation that bear directly upon the DSS and its utility. Decision theory is implemented in several ways some of which rely on such classical approaches as Bayesian statistics and others of which rely on expert systems approaches, such as rule-based methods that attempt to emulate human logic.

(3)　Computer-Based Systems
Picking and using the right set of tools is the subject of this component of DSSs. The computer instruments that are part of the decision system must have the capacity to provide the most natural and unthreatening view and control access to the operation. This includes the presentation, and switch control and other activation methodologies.

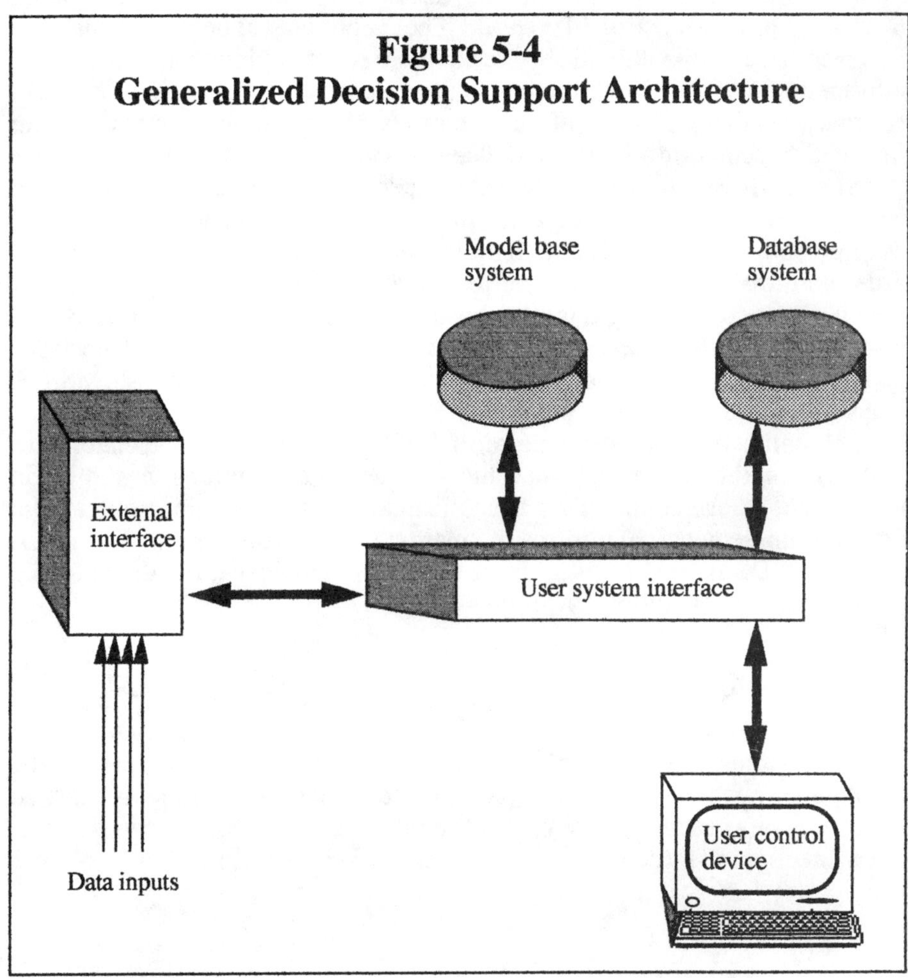

(4) Tasking Analysis

Specific problems are subjected to an analysis of their situation and specific tasks are determined based upon these analyses. The task may be related to an outage report, overload or planning need, all of which result in some task definition that awaits human review and sometimes approval. The elements of each task may be carried out by diverse elements within the system, such as a computer system, expert system, human, or switching element.

The DSS is just another step in a long line of technologies developed since World War II to cope with the needs and volume of data generated from different collection devices. The decision support systems that followed became the stimuli for the technology revolution that spawned the information age. [8] First, there was the *electronic data processing* (EDP) era that saw the collection of vast amounts of data about processes, from financial to scientific. Next, there was the *management information system* (MIS) era. During this period, a certain amount of intelligence was created by organizing the output of EDP into specific reports for management usage. The term, intelligence, may be misused, because the results were little more than reports segmented to provide insights into specific areas of the operation. Following this, there was the time when great emphasis was placed upon *management science and operations research* (MS/OR) to assist management in making choices out of the volumes of MIS reports. MS/OR helps one make decisions from highly structured information.

Finally, there is now the era of DSS. This is a set of technologies that allows the user to design his or her own solutions based upon incomplete information. This thrust provides a very fertile ground from which to implement many forms of smart systems technologies. The initial attempts at DSSs have been far from what was envisioned, but the potential is there for this evolutionary progress.

The Expert System

Expert systems are a major element of DSSs because they provide the intelligence needed for the decisions reached. They can be roughly equated to the model(s) used internally to aid the decision process. Expert systems have three main features, as shown in Figure 5-5. They are:

(1)   The Inference Engine

That software program that uses the production rules to work out the logic of a particular problem. This is done mainly through sophisticated sequential searching and comparisons using lists supplemented with external facts.

(2)   The Rule Base

That database segment that contains the rules upon which the inference engine operates to resolve the problem at hand. The rule base is essentially a model base in which resides a representation of someone's version of reality, and reveals the rule base's form of logic.

(3) The Fact Base

That portion of the database that provides the basic facts or data used by the inference engine to evaluate the rules, and therefore, to reach a conclusion. This information usually comes from external means, such as from attached or remote sensors, or from organized sources reached over communications channels.

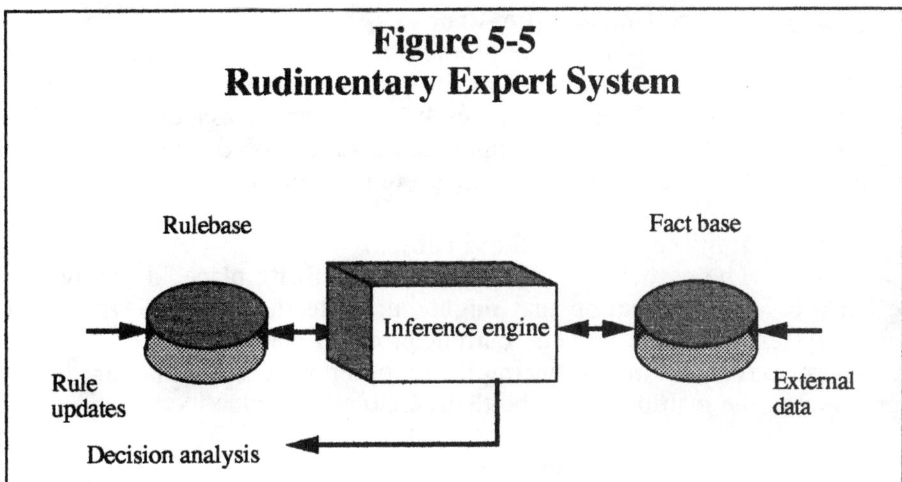

Figure 5-5
Rudimentary Expert System

Expert Maintenance Systems

Expert maintenance systems were probably the first expert systems to be developed with computer technology, and the decision support interfaces for such systems were and are, needless to say, important as well. The terms maintenance, fault control, and failure monitoring are sometimes used interchangeably to imply about the same thing, i.e., that the maintenance system has the capacity to detect, identify, and resolve system problems.

The decision support application that interfaces with the maintenance system has several requirements to satisfy in order to make it appropriate for the job at hand, and these requirements are summarized as follows [13]:

(1) Identification of the Problem

The DSS must have access to the data necessary to determine that a problem has occurred. This determination many times surfaces as a result of several indicators as opposed to a single occurrence. Problem identification may come in many forms, such as failure of a parametric test, or a logic test.

(2)   Isolation of the Problem

Determining where the problem exists is many times not an easy task. The location may be masked by several other conditions or situations that make this step a difficult one. Testing for a specific location is sometimes an effective tool.

(3)   Association of the Problem

The exposure of the problem immediately stimulates a search of the previous times or situations during which the problem may have occurred. If no such circumstances have occurred, association can be established by reference to an operational model of the system. Such events and references provide a starting point for problem attack.

(4)   Implications of the Problem

The association of the problem with its place in the system often provides information that might implicate the problem with other other system elements that the current problem may infect. The current problem may also imply or implicate other processes that are more pertinent to the malfunction or bottleneck situation.

(5)   Visualization of the Problem

A conceptualization of the problem as something that a human can appreciate is important in any system, expert or otherwise. Such a concept presentation can be provided as a graphical display or even a functioning picture of the system. The problem visualization is executed by simulation of the system functionality or processes and as a result will often suggest one or more solutions to the problem at hand.

(6)   Options for Solving the Problem

The system simulation, isolation and association are the principal vehicles for enumerating the options for problem resolution. These processes provide a list of possible options that can be implemented. In addition, the options may carry a list of caveats and degrees of difficulty that may be encountered as a result of exercising each of these options.

Performance Management Systems

The performance management system likewise has several uses for decision support. Such systems monitor operational traffic and, using decision support capabilities can anticipate bottlenecks, isolate operational trouble areas, and correlate fault messages with operational conditions. Examples of performance management systems implemented include NEMESYS and XTRAC. [4] NEMESYS is an AT&T tool developed to assist in fighting congestion in the inter-exchange carrier business. It is based upon techniques to optimize performance during overload or outage emergencies. Information on the number of and reasons for call blockages is collected and used to derive suggestions for call rerouting. XTRAC is a system that also reallocates resources based upon priority considerations.

Performance measurement and problem resolution based upon performance problems are both associated with the types of performance problems incurred within the network whether they be location specific or time based. These types of problems can be categorized as follows:

(1) Routine Capacity Overload

This is the most common problem handled by a performance management system. It occurs as a result of some cable or transmission route becoming undersized for the load placed upon it as traffic increases over that route with time. In this type of situation, the route experiences periodic durations of overload at an increasing frequency and for an increasing duration per overload period as the general loading becomes greater.

(2) Focused Overload

The focused overload results from sudden increases in local demand due to some type of emergency either manmade or natural. Examples might include severe weather or a nuclear power plant problem. The focused overload becomes most troublesome when the facility needed to complete the circuit itself is the object of the overload, such as the telephone end office, because in this case, there is little or no rerouting possible to get around the problem.

(3) General Overload

In this type of overload situation, the overload is not localized as with the focused overload. It is ubiquitous throughout the network. An example of a general overload situation would be Mother's Day or Christmas. During a general overload period, there is little rerouting and few alternatives available to resolve the problem. The problem will usually go away in a few hours. Recent catastrophes, however, have shown that general overload may not go away in a few hours. Certain local problems such as the one in New York City that affected the entire air traffic system and the software bugs that affected new telephone switches around the country are problems that may be harbingers of more dire situations in the future.

(4)   Facility Failure

Facility failure is a situation where there may be a failure of a trunk group as a result of weather problems. Here the situation resembles a focused failure except that the capacity has been reduced due to equipment failure as opposed to an increase in demand. In many such cases, the problem can be gotten around by use of spare components or switching around the problem.

Planning Management Systems

The typical planning function occurs either to support cost, schedule, or technical performance objectives. One example of a planning management system is one which operates as an expert assistant in designing networks. It is called *Designet* and acts as a configuration tool for boards, racks, cables, interfaces, etc. at each site in a network environment. [4] The DESIGNET module provides the user with many different views of the network configuration and performs sensitivity analyses by answering why and what-if questions for the operator. Another such planning system was developed at GTE Laboratories and is known as MARVEL. [9]

5.3 Interfaces

The history of human interface with computers begins with human interaction with machinery. Early computational systems used switches and hand fed tape to operate and program such machinery. Soon, the user interface developed some more complex relationships that will be discussed further below. The progression of computer input interfacing ranges from keyboard mechanical inputs to touch screens. Output interfacing has progressed from scrolling tabular presentations to three dimensional illustrations of output information. Interfaces are divided into three categories, namely formal, natural language, and graphic interface dialogues. These various dialogues are discussed below.

5.3.1   Formal Dialogue

As mentioned above, early computers were programmed and operated by switches, supplemented with tapes to input a large amount of data. Later typewriter-style keyboards were introduced to facilitate the operator's use of dedicated and time-share systems. More recently, the mouse, trackball, light pen, and touch screen have been used as input control devices. Output methods have progressed from scrolled tabular displays to icon-based presentations, to quantitative graphics, to 3-D representations of the measurements.

5.3.2　　Natural Language Dialogue

Input specifications have developed from the point where computer control required a strict protocol in order to have the user's desires obeyed by the computer. This progression has been greatly enhanced through the use of the icon, or graphic representation, to substitute for the laborious and error-prone typed input. A future innovation might be verbal control.

5.3.3　　Graphics Interface Dialogue

Icons have become the graphics tools of choice for human-machine interface. They assume the attributes of most of the previously typed file information, security, and movement specifications. Currently, icons are controlled by any of the formal dialogue methods. In the future they are likely to be controlled by touch screen or even brainwave methods.

A simplified approach to thinking of computer-user interfacing is provided in Figure 5-6 showing the basic paradigm of computer input and output control. The formal dialogue, natural language, and graphics interfaces are all involved on the input side of the system. The output side is characterized by increasingly sophisticated displays and interactions of these display buffers with external systems and databases, as well as with special display modes, such as color, scanned displays, and hardcopy generators. Input devices range from scanners to biomechanic body-sensing devices to voice activators.

While the variety of input devices is almost unlimited, the corresponding output is limited to three-dimensional and time-related imagery coupled with some form of sound accompaniment.

5.4 Planning

The development of a plan suitable for a particular DSS/EIS application begins with a rigorous elaboration of the requirements. In some cases the requirements are simply to integrate the information flow. More elaborate plans call not only for the integration of the information flow but also the ability to make unstructured requests for information and/or analyses to which the system can respond using built in intelligence, drawing from all data resources available.

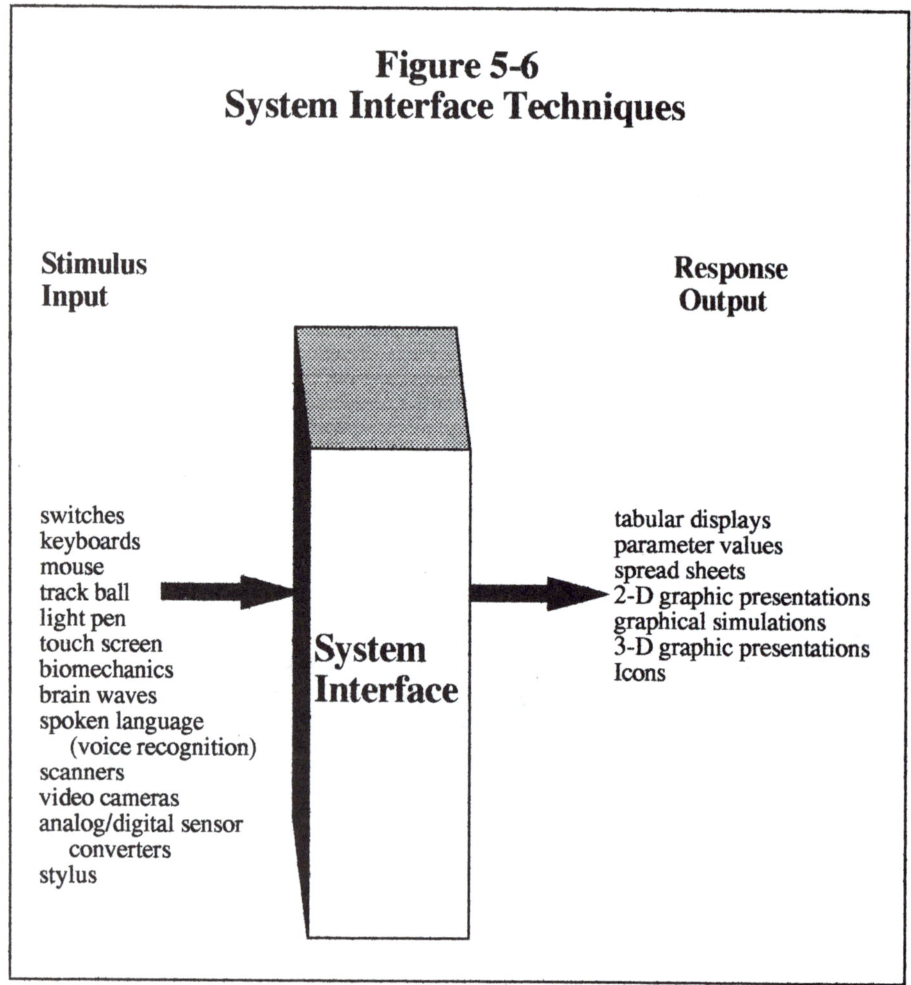

At the low end, the DSS/EIS is supported by some type of data communications capability with, at the very least, some type of E-mail facility. The high end DSS/EIS is equipped with a distributed data processing facility, fully capable of sophisticated AI-based decision support. Figure 5-7 illustrates some of the possible communications options available to the DSS/EIS developers of today.   Within a short time

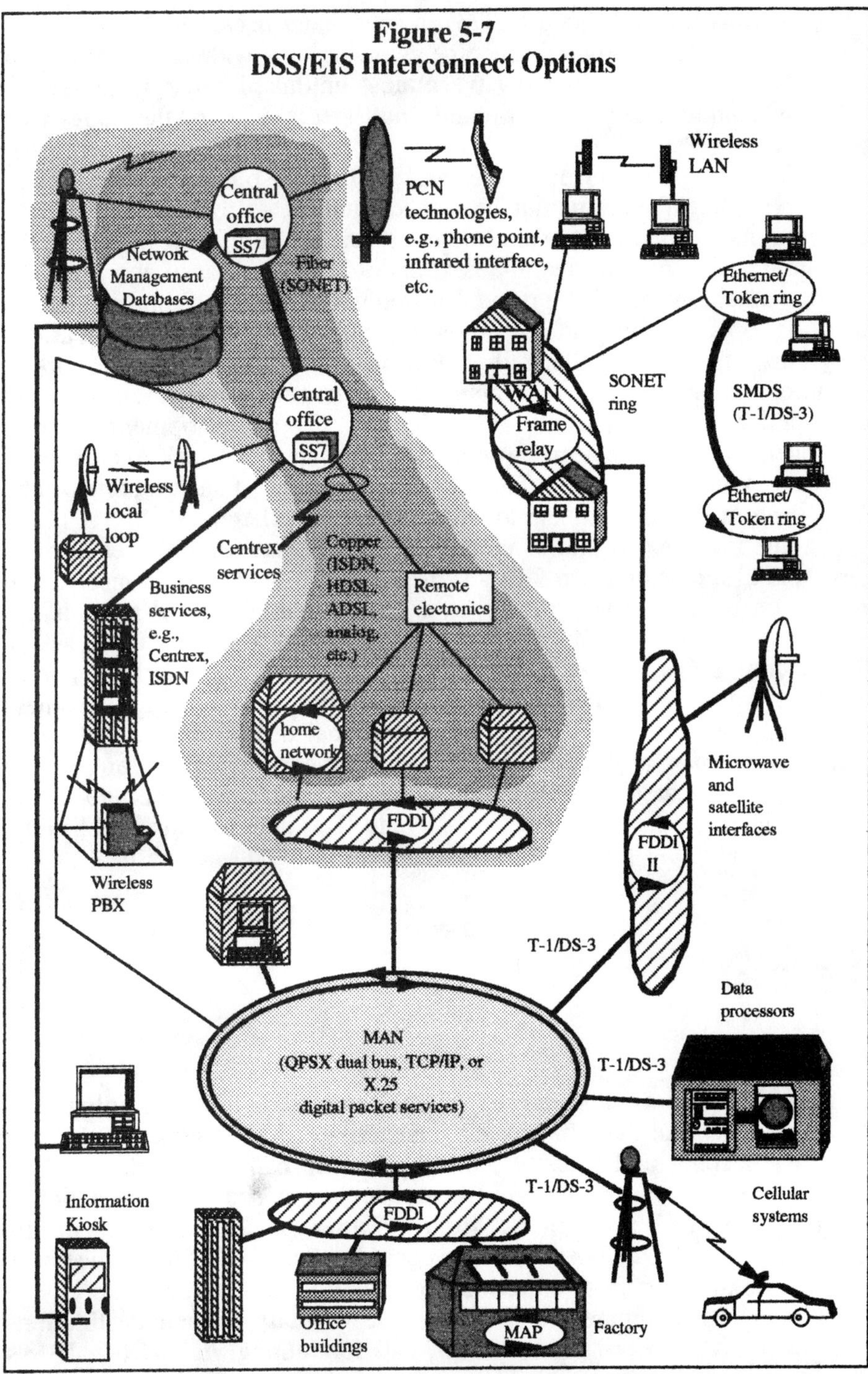

Figure 5-7
DSS/EIS Interconnect Options
Central office
SS7
Network Management Databases
PCN technologies, e.g., phone point, infrared interface, etc.
Fiber (SONET)
Wireless LAN
Ethernet/ Token ring
SONET ring
SMDS (T-1/DS-3)
WAN
Frame relay
Ethernet/ Token ring
Central office
SS7
Wireless local loop
Centrex services
Copper (ISDN, HDSL, ADSL, analog, etc.)
Remote electronics
Business services, e.g., Centrex, ISDN
home network
Microwave and satellite interfaces
FDDI
FDDI II
Wireless PBX
T-1/DS-3
Data processors
MAN
(QPSX dual bus, TCP/IP, or X.25 digital packet services)
T-1/DS-3
T-1/DS-3
Information Kiosk
FDDI
Cellular systems
Office buildings
MAP
Factory

asynchronous transfer mode (ATM), a broadband packet-based switching and transport technology will become available on a wide scale. With this capability, the developer will have almost unlimited capacity to specify different combinations of distributed, multimedia data for the target user population.

Even the low end developer and users have plenty of decisions to make regarding their communications commitments to technologies and equipments.   As the user becomes more comfortable with the communications capabilities, the next major hurdle is that of accessing the various databases and the protocol differences that invariably plague the melding of disparate data sources for access by single applications programs. The third challenge is that of managing this diverse network in terms of geography, database management, security and privileges, and communications capacity for each locale. The development of non-intelligent report generators is is next and is no small task even for the dedicated systems engineer. The types and varieties of such reports can be staggering, if the commitment to DSS/EIS is really there.

The final step is that of infusing smart systems technologies into the design and implementation. This subject will be dealt with in more detail in later chapters. For now, however, it can be said that expert systems, for example, will be pervasive in the decision maker's office of the 1990s. Some companies, such as Mrs. Field's Cookies, have taken a very aggressive position relative to DSS/EIS by setting up a "lights out" business operation that essentially makes all operational decisions with only veto power on the part of the human overseer. The current market for decision support is estimated to be about $600 million during the decade of the 1990s. The current vendors of EIS include IBM and Software Publishing, with possible likely entrants to be Borland, Lotus, and Microsoft.

5.5 Tools

The strategy being applied by most vendors in the emerging EIS marketplace is one that emphasizes connectivity first, integrated reporting facilities second, and free-style query/response capabilities third. This approach is best exemplified by the architecture posed by IBM, which is the basic architecture being put forth by other vendors as well. In this approach the host, or hosts, are the central repositories for all incoming data or file transfers from other locations, internal or external alike.

The host computers distribute requested data to their destinations that, in the case of the IBM product, call for routing via LU 6.2, IBM's communications interface for external PC/LAN configurations, 3270 terminals via their controllers, and via direct connection to DOS and OS/2 environments. The LU 6.2 interface also provides connection to various environments on the LAN, such as OS/2, DOS, and eventually Windows, UNIX, and Macintosh.

Other EIS vendors, such as Holistic Systems, have moved ahead insofar as the current wave of EIS direction is concerned. The first concentration of EIS offerings was in the area of graphically oriented, navigation interfaces that provided the user with ability to move into increasingly deeper levels of data using predefined penetration schema. These interfaces are supported by generalized as well as specialized algorithmic engines to do a lot of the preprocessing and parameter calculations. The newer wave of EIS offerings will aim at greater distributed sources and destinations and will increasingly emphasize DSS/EIS tasking at the desktop and greater simplification of the query/response process.

5.6 Case Studies

5.6.1   Frito-Lay

As the technology becomes ever more capable of supporting sophisticated information needs, the possibilities for commercial DSS/EIS approach ever more rapidly. The Frito-Lay Company is one such example of effective technology transfer from military battlefields to commercial battlefields [3]. See Figure 5-8 for a portrayal of this system. The system used by Frito-Lay, now a division of Pepsico, has evolved from an essentially manual paper operation to a paper automation system, and finally, to a full fledged decision support aid. It began with the installation of a satellite communications system capable of providing data linkages of sufficient bandwidth between field operations and regional management centers.

Second, front-line customer agents were equipped with small computers into which inventory and ordering information could be entered. These units were the key to the data collection task that is so necessary to the automation of the EIS being developed. At the end of the day, this information could then be loaded into a mainframe computer directly or remotely via modem for regional as well as local territory analysis.

Third, the data collected at the end of each period would be organized and reported directly to Frito-Lay headquarters for more immediate evaluation, decision support, and subsequent action. To tie the data together, and to put it into useful contexts, a series of report generator programs were developed aimed at answering specific questions. The human interface has become a touch screen initiation, followed by use of a group of icons that represent available reports. The managerial users now have access to easily obtainable reports that convey up-to-the-minute information in formats and through analyses that address the issues of concern to management.

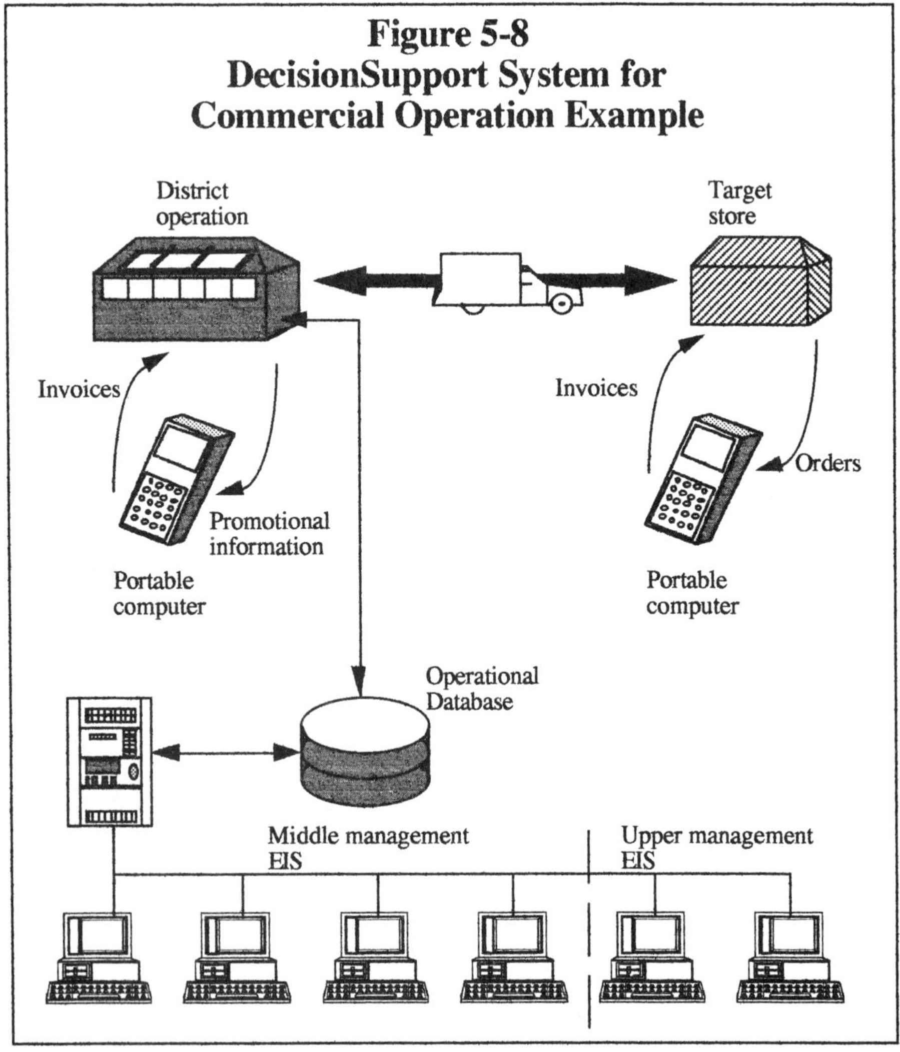

Figure 5-8
DecisionSupport System for
Commercial Operation Example

The Frito-Lay example points out some essential truths for the interested student of DSS/EIS. What Frito-Lay did was to develop a complete computer communications system from scratch that could be customized for each layer of management and for any functional interest. This computer communications system can be modified with relative ease so that new or obsolete requirements can dictate changes as needed. One of the benefits of the system was that Frito-Lay could now shift its concern from national strategy issues to what was referred to as micromarketing, i.e., marketing on a segmented local level where the revenues were made and fine tuning could be accomplished. While the Frito-Lay example per se does not relate entirely to network management, it does, in fact, illustrate how the process and requirements of DSS/EIS integrate, crystalize, and accelerate the transition of data into information and hence into decisions/actions.

Frito-Lay has developed a distributed layered EIS that provides this snack company food for action as well as thought. The system is available in at least two flavors for both the upper and middle management teams so that they can track, plan, and execute their operations more effectively. The main objective of this EIS is to carve up the massive array of constant data flowing to the operation into small enough pieces that can be digested by even the most modest of the desktop systems at the levels indicated. Each targeted display station also receives the data in a form that is concise and focused upon the concerns of that particular station user.

The company started at the bottom of the management strata and built the information system upward. The system has two types of information, that pertaining to the Frito-Lay products and that pertaining to competitors' products. The basic information gathered is obtained from the field supply trucks that collect and enter into portable computers the inventory levels at each store and order information. The systems also have the capability to print invoices for product delivered at each stop on a route. At the end of a day the portable computer is interfaced to a larger computer at a regional distribution center to which the invoices are copied, and from which pricing and promotional information is received for use on the next day's route. The timely volume of data now flowing into the company allowed upper management to be in closer touch with current operations, thus providing them with the means to avert problems as they started.

The information now available to upper management, in graphical and pictorial chart form, included reports on regional status, product status, sales status, and production and delivery status. Information on competitors is also available to company executives and is updated weekly. Direct and indirect expenses are also tracked and available for evaluation. For product promotion, the company developed a trade development system that allows sales representatives to show store managers how combinations of Frito-Lay products with other non-Frito-Lay products could stimulate sales.

Finally, the system also collects information about competitors, using the same portable computers that the delivery drivers use to track inventory and generate invoices from placed orders. Such information includes competitor pricing and presentations and is used to evaluate how the company is doing versus its competitors.

The EIS illustrated here points out several lessons that can be recapped briefly. First, the system must be implemented at the very lowest levels of the organization in order to provide worthwhile information in a timely fashion upward through the organizational structure. Second, the information provided must be presented in a way so that it can be made useful to the various management tiers requiring it, such as charts, graphs, and characterizations. Third, the information must be segmented to the maximum extent practical so that very specific answers can be extracted for very specific questions. And fourth, the data must be organized and presented differently for the different levels of management using it so that proper understanding can be maintained.

5.6.2   Quaker Oats

The Quaker Oats DSS is oriented toward marketing as opposed to the sales example cited above. The system developed is known as Mikey and basically provides several hundred users with the on-line access to Mikey and the ability to make ad hoc queries from large corporate databases. The system is PC-based, connecting hundreds of such workstations with the corporate databases via phone line connection using modems. The system is distributed, meaning that the various ad hoc queries access various databases available in order to assimilate the report requested by the user. The company DSS relies heavily upon its database management system (DBMS).

DBMSs are at the heart of any acceptable DSS. A relational system is especially important for direct queries associated with specific pieces of information. Another key issue related to the effective implementation of a DSS is ease of use. The system was therefore designed as a menu-driven operation in order to minimize any ease-of-use issues. Corollaries to this requirement are that the system must save on training time and have an integrated planning and reporting database.

The Mikey applications modules include business review, market planning, ad hoc reporting, general information queries, and utilities, such as those needed for graphical outputs. The business review module shows historical and competitive comparisons. Market planning provides optimization analyses in response to "what if" questions. For example, changes in package weights, would in turn affect variable costs to produce, that would in turn affect demand, that would affect transportation, distribution, shelf-life, storage, competitive responses, pricing, etc. The market planning application analyzes these parameters to advise the user about consequences of such moves. The ad hoc application allows the user to access his or her own set of parameters and information about market issues and to store and retrieve such information as required for future usage. A help module explains the calculations used in the various applications, and explains the menu items presented to the user. As mentioned above, utilities are provided as are database utilities that set up printer operations, convert data to graphic formats, and send data to other computer systems within the company.

5.6.3 U.S. Sprint

Sprint began its development of a DSS/EIS with a considerable structure already in place and was faced with the requirement of building over an infrastructure that was not entirely flexible. For instance, IBM PCs or PC clones and Apple Macintosh's were co-standards throughout the company. Fortunately, the developers assumed a user-needs-come-first philosophy as opposed to a technology-first position. The desktop workstations became the focus for solution development, and the mainframes became the data servers. Structural standards and interconnectivity between hardware and applications became the major objectives necessary to realize the flexibility of usage desired.

The development efforts zeroed in on the basic need to pull in data from diverse DBMSs, such as IMS, DB2, and IDMS, that was available to both PC/Mac and mainframe systems, and that could bring the data into a central repository and redistribute it to the end desktop systems. Hardware and software connectivity are key issues in such endeavors as the integration of these various separate tools.

The resultant system used a package called Focus to serve these needs. Data was brought in to the mainframes and organized as department or business views. From these views diverse ad hoc queries can be satisfied at the requesting desktop, such as business unit status reports, financial summaries, human resources information, operational details, and planning projections. Because the focus was on the desktop and customer focused, the user interface became the most important attribute of the system development effort. The decision support package available to the user is one containing an effective graphical interface, using mouse-controlled windows, icons, menus, pointers, and scroll bars.

Access is the name for the system developed and relies upon distributed functionality and reach, while maintaining centralized control over data distribution. Individual users having access to the system are granted privileges to call data types and report generators that will assemble the displays required.

5.6.4   State of Washington Office of Financial Management

The problem addressed by the financial and budget overseer for the State of Washington was essentially incompatible data sources or absence of data capture into a computer environment at all. This problem manifested itself in a lack of consistency across departments, inaccurate data collection, and delays in information availability. Resolution of the problem resulted in a common database system that was able to collect data from several other databases, and in turn could be available to a limited number of executives. The implementation of this solution also included the communications necessary to carry out this integration, and the report generators needed for the requirements at hand.

Summary:

Any hardware or software system that provides information about its operation to the user of that system is a *decision support system*. The term has grown to be somewhat more selective through restriction of its meaning to systems that provide some intelligence to, or analysis of, the raw data collected, so that informed decisions can be made about and actions taken concerning that system. The collection, organization, and even decision-oriented analysis of information pertinent to network management operations is of little value unless human concurrence and control can be exercised over that information. DSSs provide this human-oriented capability using highly interactive design tools. DSSs were first developed and implemented in numerous ways and places within the military environment over the past 50 years. More recently, the concepts involved have migrated to the commercial world for use in finance, manufacturing, service-providing, and distribution arenas, and have begun to be known as executive information systems. [12]

<u>References</u>

(1)   Bertuol, B., "Sensors as Components for Automotive Systems," *Sensors and Actuators A*, 25-27 (1991), pp 95-102.

(2)   Ericson, C., L. T. Ericson, and D. Minoli (eds.), *Expert Systems Applications in Integrated Network Management*, Artech House, Inc., Norwood, MA, 1989.

(3)   Feder, B. J., "Frito-Lay's Speedy Data Network," *New York Times*, November, 8, 1990.

(4)   Goyal, S. K., and R. W. Worrest, "Expert System Applications to Network Management," in *Expert System Applications to Telecommunications*, Liebowitz, J.(ed.), John Wiley, New York, 1988, p 1.

(5)   Harris, C. J., and I. White (eds.), *Advances in Command, Control & Communication Systems*, Peter Peregrinus Ltd., London, UK, 1987.

(6)   Henning, W., "Bus Systems," *Sensors and Actuators A*, 25-27 (1991), pp 109-113.

(7)   Hoffman, M., "Technology Profile Neural Networks," *Techmonitoring*, SRI International, July, 1991.

(8)   Hopple, G.W., *The State of the Art in Decision Support Systems*, QED Information Sciences, Inc., Wellesley MA, 1988.

(9)   Lemmon, A., *Marvel - A Knowledge-Based Planning System*, GTE Laboratories, Internal Report, 1986.

(10)   Singleton, W. T., "Man-Machine Aspects of Command and Control," in *Advances in Command, Control and Communication Systems*, Harris, C. J., and I. White (eds.), Peter Peregrinus Ltd., London, UK, 1987.

(11)   Walker, T. C., and R. K. Miller, *Expert Systems 1990: An Assessment of Technology and Applications*, SEAI Technical Publications, Madison, GA 30650.

(12)   Wallach, R. M., "The E.I.S. State," *Computer Systems News*, March 3, 1990, p 49.

(13)   White, F. E., and J. Llinas, "Data Fusion: the process of C^3I. (Command, Control, Communications and Intelligence)," *Defense Electronics*, vol. 22, p 77, June 1990.

(14)   Wilson, G. B., "Some Aspects of Data Fusion," in *Advances in Command, Control and Communication Systems*, Harris, C. J., and I. White (eds.), Peter Peregrinus Ltd., London, England, 1987.

Part

2

Smart Systems Technologies

Part 2 Highlights:

The remaining five chapters address the specific smart systems that could be associated with network management systems. The technologies discussed include expert systems, both from the standpoint of expert maintenance systems to expert performance systems. Also discussed are neural networks, starting with their biological correlates to their theory and, finally, their design and usage.

Expert systems technology is the most mature of those discussed here. Human expertise is difficult to capture and even more difficult to integrate into human-created systems. It has limited domains, or areas of coverage, and must constantly be updated when new hardware or software is introduced into operation. Its greatest advantage at present is that it is more widely understood than any of the other current intelligent technologies.

Chapter

6

Expert Systems: Human Aspects

Chapter Highlights:

Expert systems are supposed to emulate the thoughts and reasoning of human beings. Sometimes it seems as though the opposite is true. However, human experiences are ripe for capture and replication within the realm of the computer environment. The definition, categorization, and distribution of expertise is discussed in this chapter, along with a generalized description of what an expert system is, and the various issues surrounding the expert system and the humans with whom the system must interface. Probably the key topics in a chapter that discusses the human aspects of expert systems are those of knowledge engineering and natural language processing, both of which are explored here in detail. The translation of knowledge into something with which the computer can deal and utilize is the ultimate hurdle in any expert system implementation. Because the facts must be able to fit any conclusion drawn, the identification of relationships between facts and conclusions is critical to its implementation as well.

6.1 Defining the Expert

The use of expert systems in network management environments requires that such expert systems be developed by individuals who are experts themselves in the operations of the equipment and the functions performed by these systems. [2] The identification and use of the expert is crucial to the development of any expert system, and in particular, one that supports network management systems. This is so because the typical telecommunications network is usually a one-of-a-kind system that is understood by few people in the organization in which it resides.

6.1.1   Example of an Expert

One way of looking at expert systems is to say that they are nothing more than a computer-based version of the way humans look at problems. Said another way, humans have always had expert systems operating within their brains. Of course, expert systems, by definition, are those that replicate the thoughts and reasoning of humans. The point is, however, that expert systems and human reasoning are supposed to be synonymous to the extent possible given that computers are the current substitute of choice for the human brain. A very interesting example of an expert rule-based system that illustrates how we humans are supposed to employ our reasoning abilities when examining a particular problem follows. This illustration is provided to us by the great master of deductive reasoning, Sherlock Holmes. [17] The dialogue presented is from A. Conan Doyle's *The Adventures of the Dancing Men.*

"So, Watson, ...You do not propose to invest in South African Securities?"

"How on earth did you know that?" I asked.

" ... Here are the missing links of the very simple chain:

    1.   You had chalk between your left finger and thumb when you returned from the club last night.
    2.   You put chalk there when you play billiards to steady the cue.
    3.   You never play billiards except with Thurston.

4.   You told me four weeks ago that Thurston had an option on some South African property which would expire in a month, and which he desired you to share with him.

5.   Your cheque-book is locked in my drawer, and you have not asked for the key.

6.   You do not propose to invest your money in this manner."

"How absurdly simple!" I cried.

"Quite so!" he said.

Yes, how absurdly simple, when you take the time to analyze the clues. The expert can be the one with the reasoning ability, or usually, is the one with the knowledge. The above example is a perfect illustration of how apparently disjoint rules can be harnessed to reach a conclusion that actually extended the knowledge set already available. Holmes's knowledge is simply a set of observations and interactions carried through his long time association with Dr. Watson. This is exactly how the real-life expert operates, although in most cases the expert is unaware of his or her reasoning processes.

6.1.2   Characteristics of an Expert

Next, we examine what an expert is. *Experts* tend to be people that have complete knowledge of a particular application (object) or situation, including its operation, both normal and while malfunctioning The expert also knows how to manipulate the object or situation so as to suit varying operational circumstances. [7] In pursuit of the appropriate description for the typical expert, researchers have devoted considerable energy and time. This has lead to the categorization of the attributes associated with such persons. The major attributes derived from these investigative efforts are presented below.

(1)   The expert can apply expertise efficiently.

When presented with a problem lying within the expert's domain, or realm of expertise, the expert can, first of all, recognize whether the problem does, in fact, fit within his or her area of expertise, and then, having established that it does, can understand, interpret, analyze, develop, and otherwise resolve the problem with minimum effort when compared to that required of another person who is familiar with the same problem. The term "minimum effort" is somewhat misleading because it implies speed, accuracy, and all the other attributes of a quality assessment. The expert may not appear as a Mr. Spock of *Star Trek*. Many times the expert has no idea how he or she reached certain conclusions.

(2)   The expert can utilize uncertainty to resolve problems into solutions.

The expert can extract fuzzy or disjoint information from which he or she can employ plausible inferences and reasoning from these seemingly incomplete and uncertain data sets. *Fuzzy information* is that information that may be generally true, but not always. In such cases, some probability, or certainty factor, is assigned to the rule that reflects the certainty of conclusions reached as a result of data that draws upon that rule. For example, a data outage may result from one of several causes, such as transmission link failure (probability 0.8), data transmitter failure (probability 0.08), data receiver failure (probability 0.07) and other causes (0.05). Disjoint information, on the other hand, is that information which may relate to two or more rules, or which may not relate directly to any rule.

(3)   The expert can explain and justify actions related to his or her conclusions and decisions.

Given that uncertainty many times surrounds the expert's actions, the expert can assemble the subsequent ramifications and conclusions of these actions into a coherent picture from which the expert can describe a picture of the consequences and therefore the actions that must be taken to effect the solution desired. The expert's ability to perform this task may not alter the fact that the expert many times cannot relate the steps nor rules that led to the conclusions drawn and the consequences pertaining thereto. From a pure logic point of view, the expert's inability to describe the process for reaching conclusions may underly a very complex internetwork of rules and probability assessments that may need to be recursive, i.e., iterative.

(4)   The expert can share knowledge and seek out new knowledge.

Typically, the expert can also communicate sufficiently with other experts to acquire new knowledge, or to impart knowledge to those other experts. Even though the expert may not be dealing with certainty within the realm of the subject matter, there seems to be some basic level of relationship between subject matter experts. What happens is that the experts can deal with and understand uncertainty within the context of their discussions without getting frustrated or permanently confused. The sharing of mutual interests seems to be a pursuit that often manifests itself in subject-matter societies.

(5) The expert uses or modifies rules as necessary for current circumstances.

The expert, by the nature of his or her activities, is constantly establishing rules, and when these rules need to be modified, the expert breaks the rules already in force to suit the needs at hand. The expert builds a set of rules over time that are the framework for the level of expertise concerned with the subject. From time to time, the expert realizes that certain conditions may alter some of the rules. When this occurs, the rules are broken, new ones are inserted, and the rule set is reconstructed. The effective expert computer system must also be as adept at this process as the human expert, or the replication of human expertise will not work effectively. This adept-computer-expertise objective is not yet a reality. A good example of the rule modification performed by experts is when new models of the same equipment line arrive on the scene. The old rules may need to be modified to accommodate these new circumstances.

(6) The expert can determine the relevance of his or her rule set to a specific situation.

When the circumstances warrant that the relevance of the problem is not closely related to his or her knowledge library, the expert will refer the problem to another person, or state that the problem is outside their expertise. If the situation is not related to his or her realm of expertise, the expert can recognize this situation immediately. Sometimes, the issue of relevance can create new rules that can be incorporated into the expert's rule base for later usage and/or inclusion into the logic train used by the expert.

(7) The expert's knowledge degrades gradually at the periphery of his or her domain of expertise.

There is no sharp cutoff of knowledge, but rather an overlap between the particular area of expertise and related subjects. This realm of expertise is referred to as the expert's domain. It is not necessarily well defined, but tends to overlap many other domains at the fringes of the target domain. An expert in aerodynamics may find that this subject area, or domain of expertise, overlaps with other areas, such as thermodynamics, mechanical physics and fluid mechanics, to name a few.

6.1.3   Distribution of Expertise

Investigations into the general subject of expertise have yielded interesting
results. [8] An informative comparison of the relative expertise exhibited
by different groups of individuals in pursuit of their own lines of work has
been developed by researchers as they looked at several professions. In
this portrayal, five different work categories are represented, with the
interesting result being the fact that all groups have about the same level of
expertise exhibited by the same percentages of the group population. The
diversity of backgrounds is apparent, and the possible links between
various lines of work are arguably few. The correlations between each
with the others are lacking, except in the respect that each is a competitive
environment. This comparison is presented in Figure 6-1.

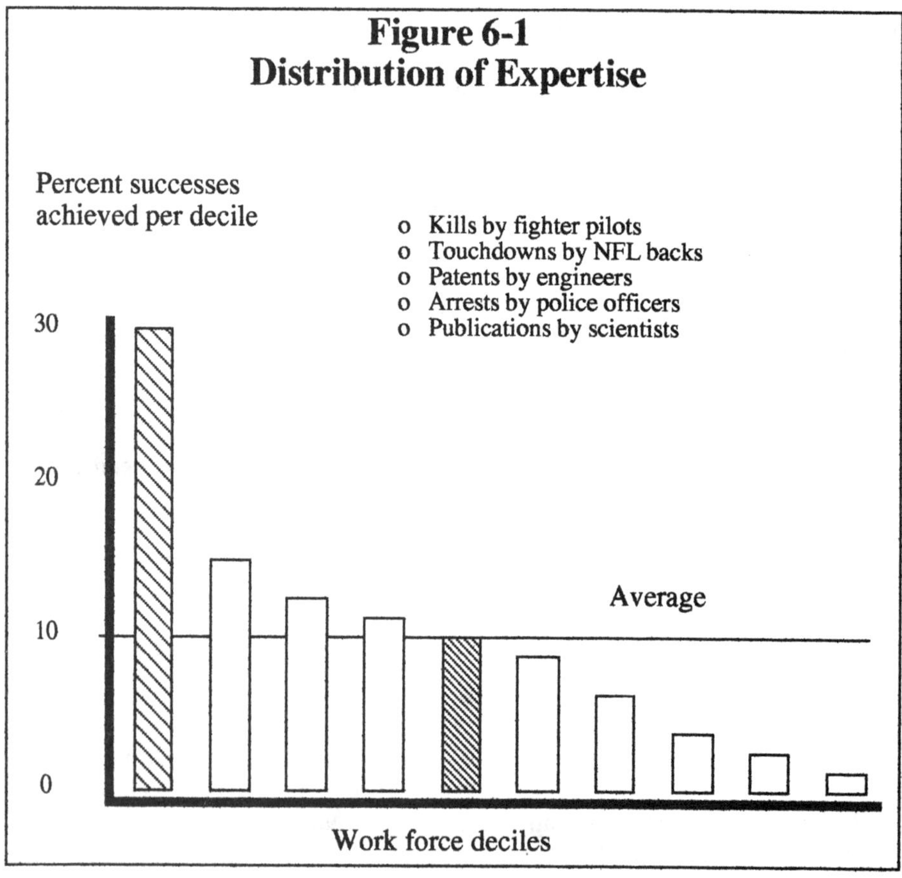

The implication is that just about all lines of work have the same percentage of experts who achieve essentially the same levels accomplishment. The illustration tends to support the old saying that 20 percent of the people in an organization do 80 percent of the work. The 50th to 60th percentile, or middle 10 percent of the organization, typically performs at the average output expected of the total group, while the top 10 percent has an output three times of what is expected of the group as an aggregate.

Apparently, no matter what the work arena involved, there are only a few, probably 10 percent, that learn the fine points of all tasks associated with the profession. This enthusiasm or sheer intellectual interest exhibited by a small percentage in each group allows this subgroup to become expert in their positions and to excel in their tasks time after time. The distribution of expertise does not follow an exponential curve, as might be expected, but rather indicates that qualified personnel meet or exceed expectations. In most situations, as represented by this bar graph, the median or top 50 percent meet or exceed average expectations, while the bottom 50 percent fall short as would be expected.

This particular example of the distribution of expertise was developed by Ed Taylor of TRW, based upon an article that appeared in the *Defense Systems Management Review*, authored by N.R. Augustine. Expertise should not be confused with intelligence, education, creativity, or just plain old long exposure to the subject. It is none of these things, but may include many or all of these elements. It appears to consist of, at least, some inner affinity to the subject such that the expert feels comfortable with the subject and shows an aptitude for handling the various sensory, motor, and/or intellectual aspects of the subject.

6.2 The Promise of Expert Systems

Expert systems have enjoyed some well-deserved praise, as well as some well-deserved criticism over the years, based in part, upon the proper or improper management of expectations about what they should achieve at any point in time. Expert systems adhere to the same rule of "garbage in, garbage out" as do any other software programs. What they can reasonably achieve is summed up in this section. [1]

First of all, expert systems are intended to improve the performance level of the specified operation. Performance, in this case, means increases in the speed and accuracy of conclusions reached in the information exchanges involved. Because expert systems contain their own rule bases, they have no trouble in remembering where they are in a "thinking" process, the problem-solving process involved, and the facts with which the process deals, to apply its rules. The accuracy of the expert system's conclusions are more a matter of their reliability or consistency than accuracy. Accuracy is equated to consistency. In the early days of expert systems development work, the emphasis was more on reliability than speed. Special purpose computers devoted to the operation of expert systems were expensive, and underpowered, and, therefore, the lack of speed hampered the widespread usage of these capabilities.

The expert system has the ability to store expertise through the rule base that it captures and encapsulates. A new job description was created in the software industry to cope with the issues surrounding this capture of expertise, and it became known as the knowledge engineer. This is a person that can interview the expert, derive his or her expertise, translate it into a form suitable for the computer's rule base capture capability to comprehend, and actually code the information into the format required of the computer. The information thus captured is stored in various forms, such as a set of rules or in frames.

Once obtained, the knowledge collected from the expert now becomes portable so that many people can have use of the information over a wide area, as well as access to the applications of that same knowledge. The computer rule base and fact base that comprise the intelligence of the system can be replicated and distributed over many workstations.

Training through direct applications of the intelligence framework with real life situations is a way of providing the user-trainee with up-to-date information so that he or she is given the most current exercising possible. This realism is most often provided by laser disks that contain the imagery necessary to replicate the problem with periodic points at which the student can match his/her level of expertise with the expert system after having learned from using the expert system provided.

Long term, the expert systems promise for the future include their usage as repositories for multi-expert knowledge where the expertise of several experts is captured and contained for subsequent usage. Additionally, the capability for extended self-learning is possible through the use of more realistic graphics and actual situations that have been collected for later student use.

Since the advent of AI as a prime influence upon the use of computers to convey words and thoughts, as well as their use to convey numbers, tabulations, and numerical narratives, there has been a considerable emphasis upon idea exchange between humans using the computer as the tool of choice. Therefore, the human aspects of any expert systems discussion can only be appreciated after understanding how such words, ideas, and thoughts are in fact conveyed from one person to another via computer technology. The sections to follow will highlight some of the major techniques used for such representations.

In actuality, the same techniques used by humans to exchange ideas and concepts are also used in the expert systems computer environment as well. Thus, as we will see shortly, the concept of the blackboard as a tool for human exchange is also used as one form of knowledge representation. The next two sections will discuss issues that deal with the human aspects of expert systems by illustrating how human thinking and intellect is used in the architecture of expert systems, as well as in understanding their limitations and problems.

6.3 General Characteristics of Expert Systems

The typical expert system has certain attributes that distinguish it from a more traditional computer program architecture. Expert systems are distinguished from standard programs by their need to have at least two information sources, namely the fact base and the database, as well as the ability to generate new facts, and the generation of multiple conclusions with varying levels of confidence. The basic structure of the typical expert system is provided below in Figure 6-2. The following are attributes and general characteristics of such a typical expert system. [3]

(1) Application Domain
Expert systems operate on rather narrow real world problems, and as a result have restricted domains of expertise. Part of the reason is because the expertise of the sources of knowledge are fairly narrow of themselves. Another aspect of this restricted scope is that the degree of expertise is somehow related to the breadth of the problem. In other words, the narrower the range of knowledge, the deeper the analysis of the problem.

(2) Approach
Expert systems operate using different types of strategies, some of which include heuristic methods, multi-pathing, and rule-based searches. Additional information repositories include the fact bases needed to judge the implications of the rules. Other approaches used involve knowledge engineering, data driven and/or goal oriented control, special purpose software, e.g., the LISP language, and in some cases special purpose hardware, although general purpose computers are increasingly capable of exercising A.I.-based programs. Multipathing involves reaching

conclusions by any of several different approaches. For instance, a conclusion about data loss may be reached by prior knowledge about a specific transmission loop, or it may be reached because general knowledge about data dropouts indicates that the usual cause in such cases is the transmission system.

(3)   Development

The typical expert system uses a special development support software package, but direct implementation through such A.I. packages as LISP is also possible in some instances. The expert system development is also characterized by the incremental and evolutionary step-by-step development activity, where the knowledge base is built up over time and modified as changes or new information becomes available to the user. The system must be supplied with current facts, as well as the factbase, and a set of rules, and the rulebase, sufficiently complete so as not to make the expert system conclusions superfluous. The general development requirements for such a system include: (1) examining facts for out-of-range conditions, (2) searching and finding rules that account for these range conditions, (3) assessing the probabilities for each of these rules, and (4) assigning a conclusion with some appropriate probability of assurance based upon all conclusions that could apply.

(4)   Evaluation

The expert system is evaluated in a qualitative sense. Since there are no rights or wrongs, the expert has to be the evaluator of what is sufficient or insufficient in the expert system operation. This is many times difficult, because the rules may suffer from being disconnected from the possible conclusions.

The user interface may take any of several forms, such as frame-based and natural language. [2] The control structure performs the analytical and logical manipulations. Such manipulations involve rule interpretations, applications of facts where appropriate or necessary to make interpretations, making inferences based upon the facts and rule applications. The fact and rule bases are the information repositories. The factbase is analogous to a database, and the rule base is an assembly of the rules that might be exercised. The factbase may also be refreshed in realtime from the external world, or be entirely internal, or be a combination of both.

The inference cycle is illustrated in Figure 6-3. Here the method of drawing conclusions and maneuvering through the various rules involved is indicated. The critical issue involved with the inference process is the interpretation of the rules executed. Faulty logic here makes for a weak system implementation.

6.4 General Issues Associated with Expert Systems

6.4.1   Limitations and Problems of Expert Systems

The overall problems associated with expert systems are somewhat evident. First of all, each expert system, of necessity, is rather limited in its domain,

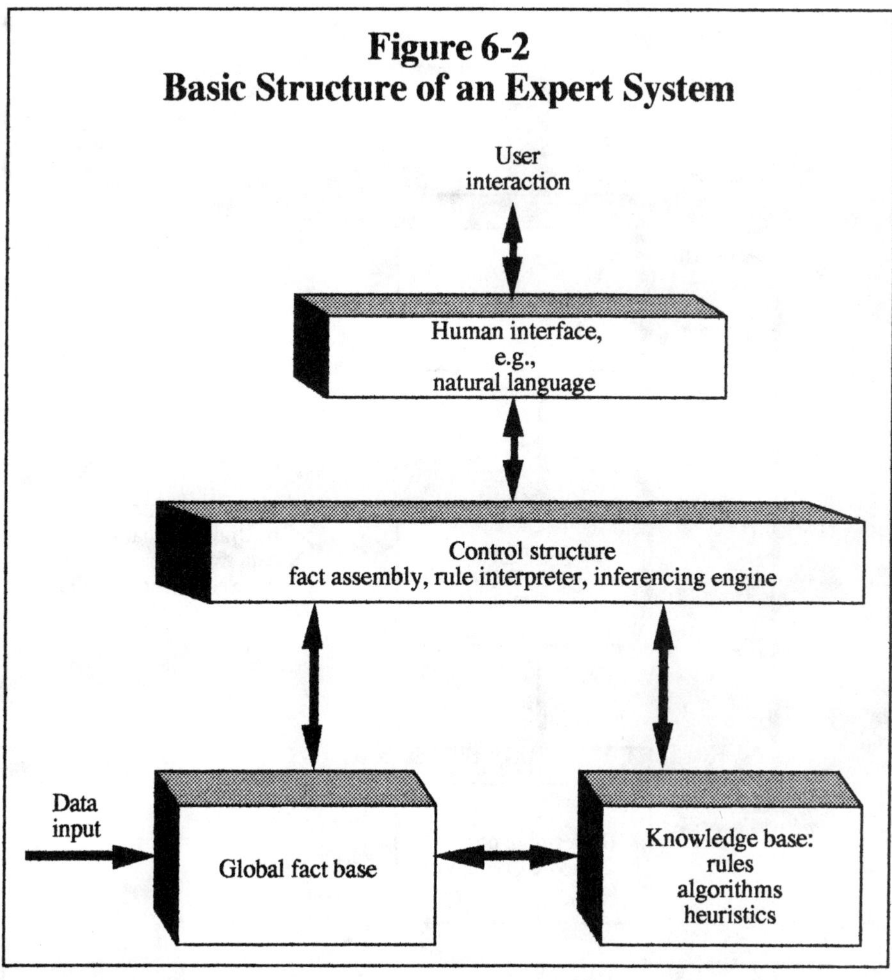

Figure 6-2
Basic Structure of an Expert System

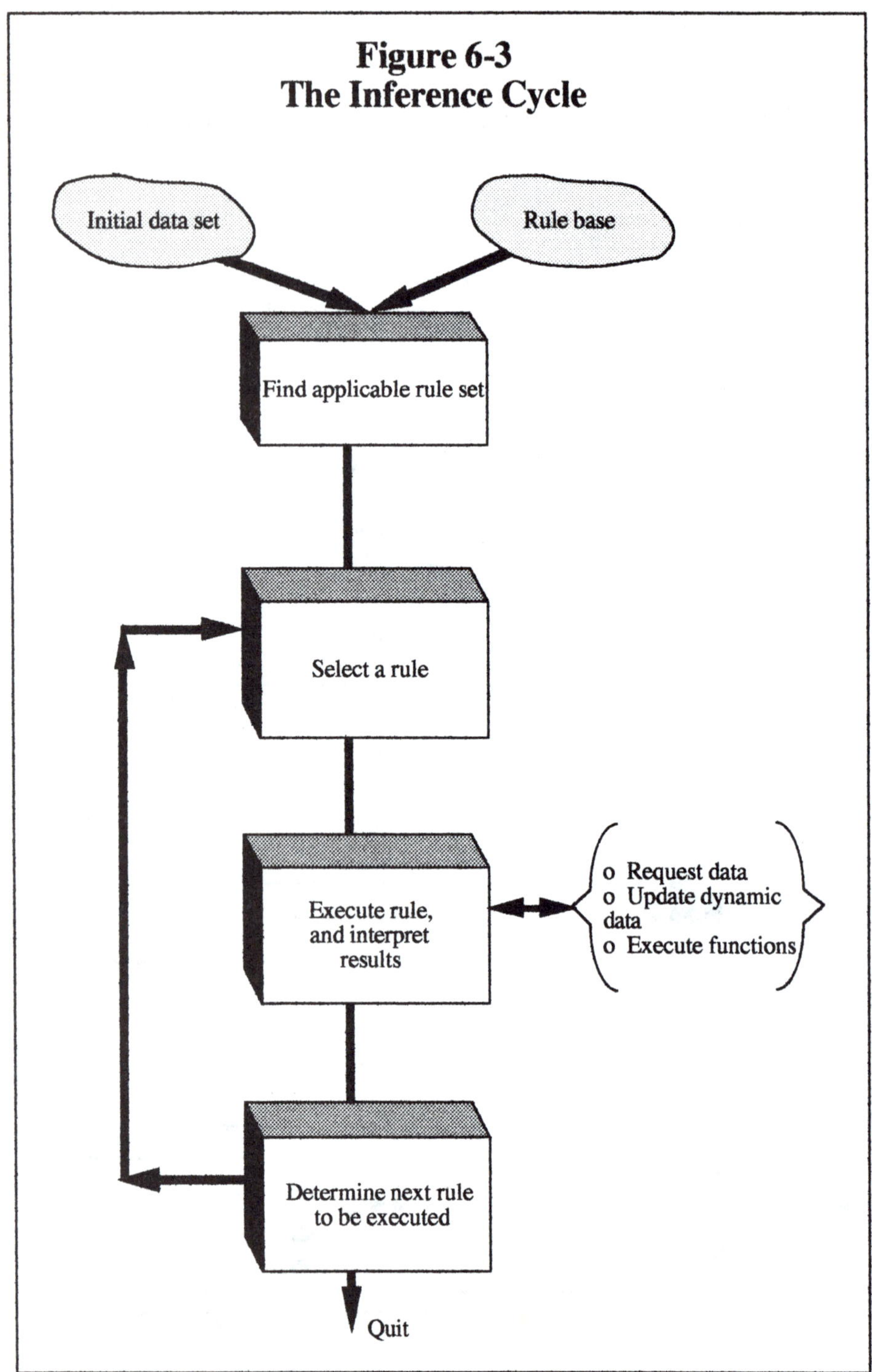
Figure 6-3
The Inference Cycle
Initial data set
Rule base
Find applicable rule set
Select a rule
Execute rule, and interpret results
o Request data
o Update dynamic data
o Execute functions
Determine next rule to be executed
Quit

or scope of knowledge. [4] They are designed and built to parallel human expertise, and since human areas of expertise tend to be limited, so too are software expert systems. Second, the implementation tends to be very laborious and time consuming, thus making the implementation very expensive. The capture and packaging of human experience is not easily translated from human intellect to a computer environment. The validity of the resulting expert system is only as good as the expert chosen to replicate the possible knowledge in that domain. In other words, the *knowledge czar* chosen picked must be a good choice.

The problems involved with expert systems can be many, but the most common are: (1) the engineering of the knowledge acquisition process; (2) the shortage of knowledge engineers; (3) lack of rapid prototype development tools; (4) search strategies are not adequate for the large and complex problems of today; (5) the representation of knowledge cannot be extracted rapidly for use in the expert systems engines; (6) the current line of machinery is inadequate in speed and processing capability for the sophisticated needs of today and tomorrow; (7) the validation of decisions and explanation of such decisions falls short of current and projected needs; (8) the typical expert system is inadequate in adaptability and at domain boundaries; (9) response times are too slow; and (10) the human interfaces for expert systems are inadequate. [5]

6.4.2   Knowledge Representation Techniques

The expert system of today requires not only faster and more flexible tools, but also new thinking as to how information, rules, and incomplete knowledge (guesses) can be injected into expert systems design. For instance, knowledge representation comes in many forms, but none of these are entirely adequate for all needs. Because humans are flexible in their thinking, they can use many paradigms for organizing and representing knowledge. Indicated below are some of the more popular examples of knowledge representation mechanisms.

(1) Production Rules

Popularly known as if-then-else rules, *production rules* are a combination of an antecedent and a consequence. They provide an assertion given that certain conditions exist. They are commonly referred to as production rules because of their action-oriented nature, i.e., they provide results. They provide the basis for decision making, given the antecedents provided. The use of if-then-else rules for computational suitability was first proposed by Emil Post while working at The City College of New York in 1943. [2] This approach was later extended for use in studies of human intelligence as well.

In a network management implementation of a set of production rules, the paradigm would encompass the physical and functional aspects of the telecommunications system. Examples of such rule sets are: If input switch set to X, then select output port B, or if input switch set to Q, then select output port K. The challenges of using production rules include identifying enough of the problem to allow a search for the applicable rules, and locating the applicable rules in the rule base.

The use of production rules is the most familiar and common approach to applying knowledge representation to problems.

(2) Networks

Networks are the result of a description of a hierarchical relationship between general objects and ever increasingly specific instances of those objects. The increasingly specific instances of those objects provide the action routes of the outcomes, because the action is the pursuit of the specific cases involved. These types of networks include transition networks, semantic networks, and procedural networks. Examples of knowledge networks range from taxonomies to classification systems.

(3) Frames

Frames are representation methods that encapsulate multiple attributes of a single object, e.g., data structures that include both declarative as well as procedural information about an object. Frame-based information is stored as pre-defined internal relations. Frames are like still pictures, each of which contains its own features and attributes.

(4) Scripts

A group of procedures or habitually used actions is known as a script. Such procedures are goal-oriented and represent a means to an end. Scripts are similar to written procedures where the conclusions are based upon the completeness and accuracy of the script provided.

(5) Blackboards

Blackboards are an intermediate domain in which experts can exchange knowledge. It is the communications medium between the competing knowledge sources, that are in reality competing experts, that share this multilevel representation of an event.

(6) Direct

Analogies, or analogical representations of a set of events can be represented by schemes, such as maps, models, diagrams, music, pictures, etc., that portray the object(s) desired.

6.4.3   Examples of Knowledge Representation

Rule-Based Knowledge Representation

Production rules are the most common representation technique in use today. They have some drawbacks, however, that are due mainly to the time-consuming nature of the translation of knowledge into a machine-usable form, and the structured nature of their form. A brief example will illustrate these drawbacks. Suppose that we are designing an expert system intended to isolate faults within a packet management system. Using the rule-based knowledge representation approach, we must exercise our knowledge engineering expertise to develop a rule for each and every fault isolation circumstance that we would anticipate. One isolation rule out of the ensemble that would have to be developed might be represented as follows.

If   Two packets arrive at relay point A in error within 2 seconds of each other;

And   The two packets use the same incoming circuit;

And   The two packets use PAD (packet assembler / disassembler Number 1;

Then   The fault is with PAD Number 1 (with certainty of 0.9), or the incoming circuit (with certainty of 0.1).

Notice that the engineer charged with the responsibility of developing the expert system for tracking down errors and diagnosing such problems must meticulously analyze all possible circumstances and provide diagnoses for all potential problems. This requires that the analyst, or knowledge engineer, must be aware of the physical structure of the system as well as its functional operation. A very simple trouble-shooting system with three levels of conditions can result in many levels of different outcomes.   Typically, if the conditions are bimodal, three different conditions can result in eight outcomes, i.e., 2^3 possibilities.

Script-Based Knowledge Representation

Script-based representations of knowledge portray exactly what one might expect. They provide the setting, narrative, and scenes associated with a situation in which information is to be conveyed. Indicated below is a brief example designed to acquaint the reader with the basic idea behind the technique.

Script	Assemble packets
Setting	Normal communications operations
Scene 1	Packet information is presented to the PAD interface
Scene 2	PAD performs checks for packet length, address(s), and routing information
Scene 3	PAD assembles packet with proper check codes, packet sequence number, etc.
Scene 4	PAD releases packet to circuit

The script approach to knowledge representation provides a step-by-step look at what should occur with each outcome. An abnormal event will single out a particular scene for scrutiny. For instance, if a garbled sequence number is inserted into the packet sequence number field, then the events associated with scene 3 are examined for possible action.

Semantic Network

A semantic network is akin to a functional flow breakdown. The subject is broken down to a finer and finer level of granularity, so that the subject is more and more specific. Everyday examples of such devices are organizational charts or troubleshooting diagrams found in appliance owner's manuals. Semantic networks are also working examples of inductive reasoning where specific circumstances are examined so as to imply a general solution. Shown below in Figure 6-4 is one such brief example.

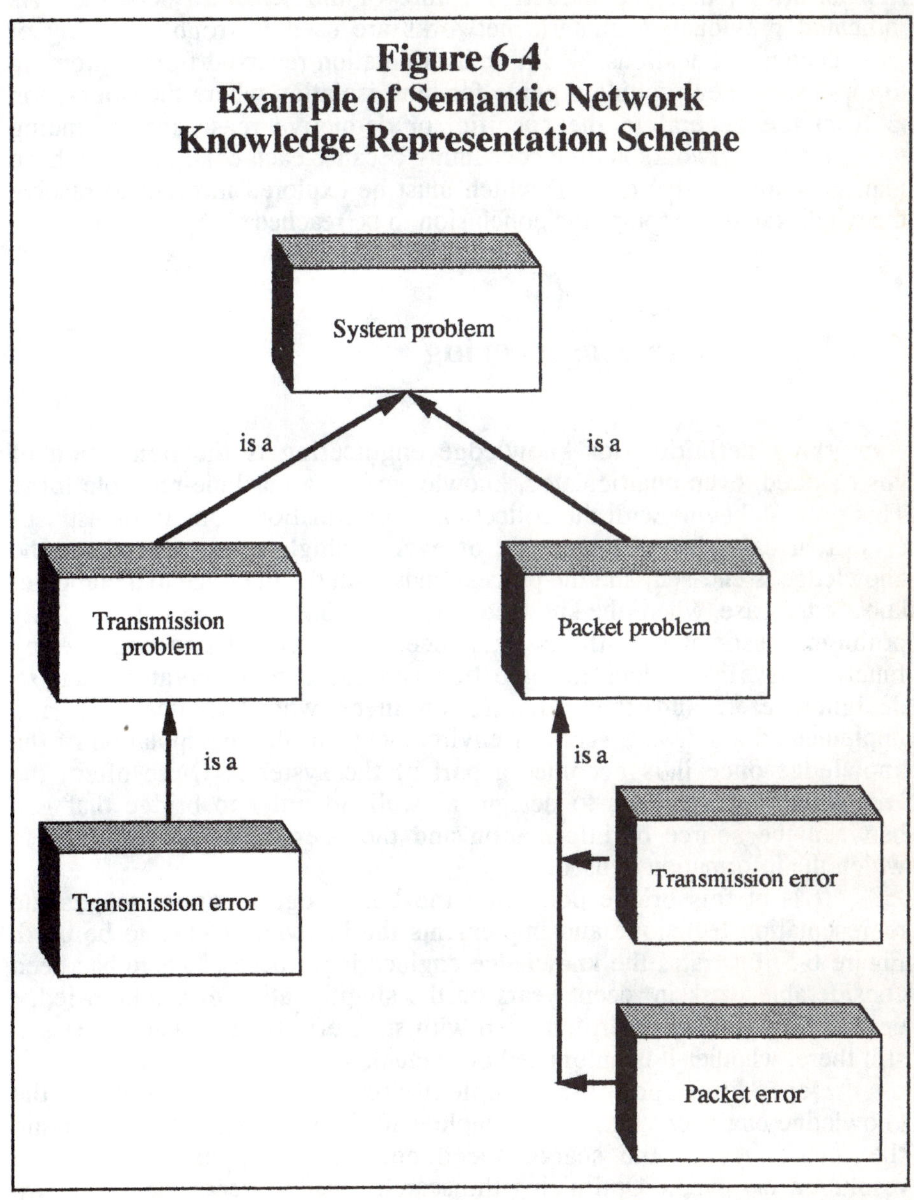

Figure 6-4
Example of Semantic Network
Knowledge Representation Scheme

Here system problems break down into either transmission or packet problems. Each of these, in turn, breaks down into either transmission or packet assembly or disassembly errors. The symptoms, however, occur first as errors, thus the inductive nature of the semantic network. As indicated previously, semantic networks are used in troubleshooting or maintenance applications. With their orientation reversed from bottom up to top down, they provide a guide for fault isolation where the orientation is from the general to the specific, or deductive reasoning. Tracing through such networks is time consuming because each condition may have many resultant paths, each of which must be explored in order to resolve the condition and pinpoint the conclusion to be reached.

6.5 Knowledge Engineering

A working definition for knowledge engineering is the translation of unstructured, even unarticulated, knowledge into a machine-readable form. This process begins with the collection of information from an industry or a craft, usually involving people or even a single person in whom the knowledge is encased, and the process ends with the interface to a computer knowledge base, where the knowledge resides while awaiting use in solving problems associated with its existence. The knowledge engineering function usually bridges the gap between the expert operator, analyst, designer, etc., and the software engineer who has designed and implemented a software shell, or environment, for the manipulation of the knowledge once it is becomes a part of the system. Quite often, the knowledge engineer has to design, as well, in order to bridge that gap between the source of information and the operational environment in which the information is used.

It is at this bridge point that the knowledge engineer selects the representation technique and implements the knowledge base to be used. Figure 6-5 illustrates the knowledge engineering process. There has been considerable work in recent years on the simplification of the knowledge engineering process itself, but even with such efforts, the basic process is still there, whether it is automated or manual.

Depending upon the complexity of the search problem, the knowledge engineer will have to implement the solution, keeping in mind the importance of the search speed needed to support the decision resolution involved. Optimizing the search speed is dependent upon two major factors, one being the size of the search problem itself, and the other being the resources available in the system to effect the problem resolution.

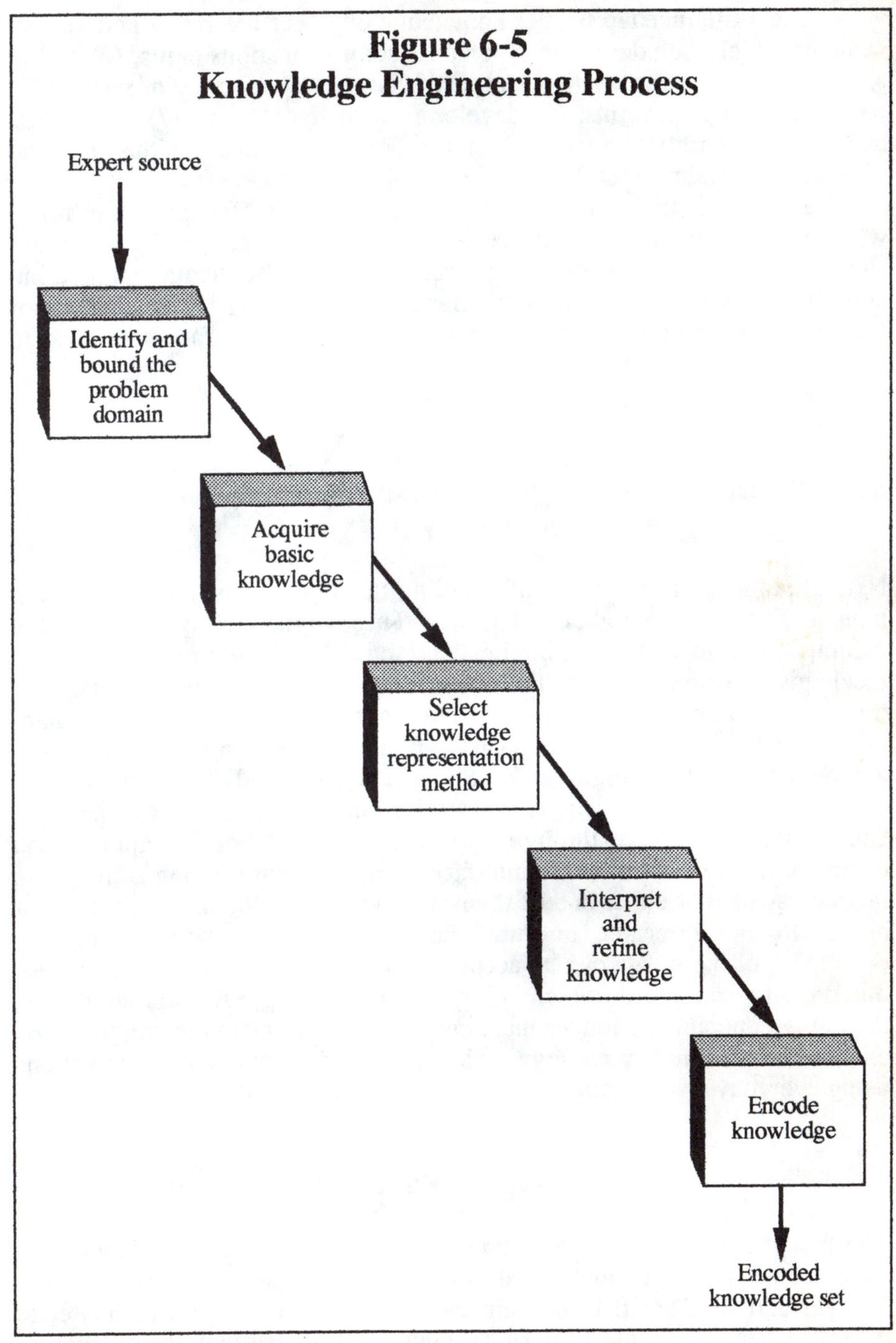

Figure 6-5
Knowledge Engineering Process
Expert source
Identify and bound the problem domain
Acquire basic knowledge
Select knowledge representation method
Interpret and refine knowledge
Encode knowledge
Encoded knowledge set

The skills needed by the knowledge engineer are many and varied, some of which include: (1) interpersonal communications skills, (2) ability to summarize, (3) explanation and justification, (4) assembly of results, (5) recognition of constraints, (6) development of hypotheses, (7) assessment and use of certainty factors, (8) organization and control of situations, and (9) computer technical skills.

Thus, the translation of knowledge to machine-readable form, whether it be automated or exercised by another human, is an important step in the development of expert systems. This knowledge engineering activity can be as complex as the human intellect, but the limitations of humans will continue to make it an imperfect process for some time to come.

6.6 Natural Language Processing

Natural language is a method of handling the signs and symbols by which humans exchange thoughts and ideas. These signs and symbols are the natural language of human intellect. Natural language processing is the mechanism through which the ideas and thoughts of one human are transmitted to another human. The other approach to interface with man-made systems is via formal languages. Examples of such formal languages are the programming languages BASIC and Ada, or CAD/CAM languages.

The normal processing chain for information collection and usage is from sensor and source through whatever transformation is required, and ending at the human-computer interface point. This processing activity can be implemented at either end of the processing chain, the incoming data, or at the human interface output. For example, the natural language processing of input data can be accomplished by any number of techniques, one popular example of which is key word searching. At the output side, the human operator no longer must encode instructions to the computer via esoteric formats and symbology. The operator now may enter instructions using everyday words and formats, or very near equivalents.

6.6.1   Objectives of Natural Language Processing

The reasons for studying computational linguistics, or natural languages, have been developing since the dawn of language itself. In 1946, Warren Weaver and A. Donald Booth suggested the use of a digital computer to perform machine translation of language. The original vision was to develop a mechanical device that could read documents in one language and to have the same document reproduced in another language. Indicated below are some of the objectives for natural language processing.

(1)　Document Translation

Convert a document from one language into a document of another language while retaining the meanings of the original or source document.

(2)　Document Understanding

Decode and associate a document's contents with other information within a large framework of knowledge. The process dealing with the document might generate document abstracts, answer specific questions about the document, or alert previously identified individuals about the document.

(3)　Document Generation

Formalized descriptions may be translated into natural language via a machine transformation capability. Physical design will, at some point, have its operational characteristics understood and formulated into normal everyday language. Another form of document generation is hypertext where parts of several documents are assimilated into a single document.

6.6.2　Natural Language Processing Definitions

The issues of natural language processing involve the terminology of the study. Provided below are some of the more important definitions of terms used in this work and some of the implications associated with these definitions will be outlined to show the challenges to be faced.

(1)　Grammar

The structural scheme for assembling the language is known as the language grammar. The structural schemes of any language have numerous exceptions that tend to make machine reading or machine generation of that language difficult. The handling of grammar, therefore, is a major obstacle to the implementation of machine language processing; however considerable progress is being made in this area.

(2)　Parsing

Parsing is the use of rules or predetermined requirements to identify certain features or language parts of the input language stream. It is a means of understanding control aspects of the word stream for the purpose of setting up or taking needed actions.

(3)   Derivation Tree

A derivation tree is a word structure that depicts the relationships between the various words in a sentence.   It is similar to sentence diagramming, a technique used in high school English to illustrate the parts of speech and their interactions.

(4)   Syntax

The rules that govern the formats of phrases and sentences are known as syntax. It is syntax that provides the basic structure of language.

(5)   Semantics

Semantics is the study of the meanings of the words and expressions of a particular language.

(6)   Text Generation

This is the creation of sentences by a computer program. This activity requires that the syntax, semantics, and grammar be incorporated into the system.

(7)   Machine Translation

The translation of a document from one language into another language is known as machine translation. This act requires that the system be acquainted with both languages in terms of their syntax, grammar, etc. In certain literature where there are well-structured words and phrases, the translation accuracy is about 85 percent. Examples of such literature would include engineering and scientific journals. For less structured documents, such as newspapers, where the thought process begins to affect the translation accuracy, the accuracy drops to 50 percent or less. For poems, or other pure literature, the idiosyncrasies of the language come into play, and the accuracy drops even lower to about 30 percent.

6.6.3   Approaches to Natural Language Processing

There is a progression of ways by which natural language processing can be accomplished. The major processing steps are discussed briefly below.

(1)   Key Word Matching Approach

Incoming sentences are checked for words or patterns that have been previously declared to be the search objects. Depending upon the application program running the required key words or patterns, the search objects may be together or separated throughout the incoming word stream.

(2) Text-Based Approach

With this approach, the incoming word stream is stored in a database and retrieved according to a some indexing scheme. Indexing schemes operate by assigning indices to the material and searching for a specific instance of an index by scanning through the material for that instance. The scanning process may be a direct lookup, or some type of coded lookup algorithm.

(3) Limited Logic Systems Approach

In this type of natural language processing, the material is again stored in a database and tagged with some type of identifiable notation. Programs translate retrieved sentences into an internal format, and logic rules are implemented to form inferences about the processed word stream.

(4) Knowledge-Based Systems Approach

Knowledge about the domain of the word stream is stored in a knowledge base. This knowledge base has the logic, semantics, frame data, etc. that is used to understand the language of the incoming document stream. Thus, this type of processing approach accepts narratives as inputs, can paraphrase the document, can answer questions about the document, and can make inferences about the information provided in the document.

6.7 AI System Development

The various elements that comprise the human thinking and reasoning processes must now be collected and assembled into a computer system in order for it to be used as an AI environment. This activity will be discussed here. The basic process of developing an A.I. system is illustrated in Figure 6-6, and following discussions will relate to the order shown here.

6.7.1   Defining Goals

Defining goals in this context means that the designer has to know or be told what the owner wants to accomplish, i.e., his objectives. These objectives are usually defined in some statement of requirements that boils down to describing how the various elements of the system interact with each other in various situations. Finding out what the owner wants is many times much easier said than done, but essentially consists of answering the following questions:

(1)   Problem Solution

What sort of problem is to be solved, i.e., what disciplines are required for the implementation of the solution, and what are the subjects to be attacked?

(2)   Problem Boundaries

What are the boundaries of the problem space, i.e., what areas does it cover, and what are the constraints of the investigation?

(3)   Accuracies Required

What level of detail is expected, i.e., what level of ambiguity is tolerated, and what resolution is required? In expert systems engineering there are several criteria to be dealt with, among which are fuzzy logic, statistical inferencing, and frames of reference.

(4)   Success Criteria

How is success to be measured, i.e., what degree of parameter identification, or quantification can result in a claim of success?

6.7.2   Define the Approach to the Solution

The solution approach involves the description of the system architecture. In order to reach this point, the functional requirements must be derived based upon trade-offs between user needs, technology availability, and business and financial constraints. Once the architecture is determined, it must be subjected to costing, designed, and productized, i.e., organized for delivery as a product, as necessary for use in its intended mode.

The expert system element of the system is derived from a variety of considerations. If the system attempts to address a large enough domain, then the rule-based approach may be inadequate from the standpoint of generating a timely set of conclusions, or may require an inordinately large rule base and attendant storage requirements for same. The frame-based approach may suffer from inadequate amounts of information needed to maintain its views of specific objects. Thus, the expert system determined to be necessary to satisfy whatever requirement will likely be a compromise of several conflicting issues, just as is the case with any other technology question.

6.7.3   Fact Definition

Facts are an essential ingredient to any intelligent system implementation. They may come into the system in any number of ways, for instance, by manual input via a keyboard, or by way of an automatic and continuous digital sensing function that regularly samples the external environment, or through any number of other means in between. The overall capability profile for distributed data bases is as follows:

(1)   Data Dictionary
There is a data dictionary that defines the formats, scaling and units for data collected and stored in the database.

(2)   Database Servicing
The database contains priorities for the update cycles and queries made to the database.

(3)   Database Capabilities
The database has facilities for the capture and retrieval of facts.

(4)   External Requests
Each external request is retained for later update purposes.

(5)   Data Reconciliation
Each external update request is retained for reconciliation at each update cycle time.

6.7.4   Data Acquisition

The data acquisition phase is the point at which information is gathered so that the facts, previously defined in the rule definition step, can be evaluated. In this phase, the responder or environment will be tasked to provide information that can then be structured to answer the implied questions at hand. This process can operate in a manner similar to the following example.

Suppose the definition phase yielded a set of questions to which positive responses would identify a duck. The fact has already been established that we are dealing with an animal. These questions might be stated as follows:

(1)   Does the animal waddle while walking?
(2)   Does the animal have webbed feet?
(3)   Does the animal have a bill?

(4)   Does the animal have feathers?
(5)   Does the animal quack?

A weighting factor is assigned to this and other identification questions so that the resulting categorization can be evaluated to yield an answer with some level of confidence. In this case, the weighting factor will score each of the "yes" answers with a 1. For this example, a score of 5 is required to assert with a confidence level of, say 0.95, that the animal being discussed is a duck. A score of 4 might yield only a 0.15 confidence level that the animal is a duck. A score of 3 might negate any chance that the animal is a duck.

We need rules to apply the facts obtained to the situations that arise. In this particular situation, the rules help us to evaluate whether data collected about some object supports its being classified as a duck. A score of 5 is considered a threshold value such that any result below this will question the hypothesis that we have a duck on our hands. Rules used to evaluate this situation might be declared as follows:

If the score obtained is 5, then the animal is a duck with a confidence level of 0.95.

If the score is 4 or less, then the animal is likely to be another type of animal.

Thus, the method outlined above is directed at providing a conclusion based upon previously established facts. This is known as forward chaining. A situation in which we have established our conclusion and are looking for corroborating evidence, or facts, is known as backward chaining. Such a situation could be approached by having established a proposition, such as Johnny is a genius. This proposition was based upon several facts, among which was the fact that he has scored 150 on an IQ test. To gain further assurance that this conclusion is correct, we might look for other supporting evidence, as, for example, the fact that he can play 10 games of chess simultaneously while blindfolded.

The danger in any logic exercise such as this, of course, is that the facts must relate to the conclusion; otherwise we will be guilty of what logisticians call *begging the question*. Begging the question is assuming as fact something that has not been proven. Here, as in any logic association found in an expert system, the facts must relate appropriately to the possible conclusions, otherwise, the conclusions reached are irrelevant to the facts, and the system is useless.

Summary:

Any hardware or software system that provides information about its operation to the user of that system, is a *decision support system.* The term applies to systems that provide some intelligence to, or analysis of, the raw data collected, so that informed decisions can be made about and actions taken concerning that system. The collection, organization, and even decision-oriented analysis of information pertinent to network management operations is of little value unless human concurrence and control can be exercised over that information. Decision support systems provide this human-oriented capability using highly interactive design tools. Decision support systems were first developed and implemented in numerous ways and places within the military environment over the past 50 years. More recently, the concepts involved have been migrated to the commercial world for use in finance, manufacturing, service provision, and distribution arenas, and have begun to be known as executive information systems. [6]

<u>References</u>

(1)   Goyal, S. K., and R. W. Worrest, "Expert System Applications to Network Management," in *Expert System Applications to Telecommunications*, Liebowitz, J. (ed.), John Wiley, New York, 1988, p 1.

(2)   Hall, D., & J. Llinas, *Advanced Techniques for Data Fusion*, Seminar for the Education Foundation of the Data Processing Management Association, 1986.

(3)   Kiggans, R., *Expert Systems*, Education Foundation of the Data Processing Management Association, 1986.

(4)   Lamont, G.B., *Knowledge-Based Software Development*, Education Foundation of the Data Processing Management Association, 1985.

(5)   Walker, T. C., and R. K. Miller, *Expert Systems 1990: An Assessment of Technology and Applications*, SEAI Technical Publications, Madison, GA 30650.

(6)   Wallach, R. M., "The E.I.S. State," *Computer Systems News*, March 3, 1990, p 49.

(7)   White, F. E., and J. Llinas, "Data Fusion: the Process of C^3I (Command, Control, Communications and Intelligence)," *Defense Electronics* V 22, p 77, June, 1990.

(8)   Wilson, G. B., "Some Aspects of Data Fusion," in *Advances in Command, Control and Communication Systems*, Harris, C. J., and I. White (eds.), Peter Peregrinus Ltd., London, England, 1987.

Chapter

7

Expert Systems: Maintenance Systems

Chapter Highlights:

Expert systems have been heralded as the basic tool of the 1990s. Their appeal as a means of replicating human judgment, with the attendant cost savings and consistency of application, has put expert systems high on the agenda of many developers and users alike. Probably the most directly applicable use of expert systems to network management has been in the realm of maintenance operations. Operations and maintenance are the two most critical areas of current-day interests in network management, and as such their relationship to AI and expert systems influence, in particular, cannot be overemphasized. Provided in this chapter are the fundamentals of this critical and cost-saving technology of this decade if not beyond.

7.1 History

Any discussion of expert maintenance systems must clearly begin with a discussion of AI in general. The origin of interest in AI, as well as robotics, goes back to the time when ancient Greece was the dominant force in the world. Greek myth relates the story of Pygmalion, in which a Cypriot king had an ivory figure of a female carved. After completion of the work, the king falls in love with the statue. [8] Aphrodite, the goddess of love, sees his anguish from her vantage point on Mt. Olympus, takes pity on the king, and grants his wish to bring the statue to life. This was probably the first recorded discussion of a surrogate robot.

Since then, less romantic but more tangible progress toward realizing the "dreams" of cybernetics, the umbrella science for AI and robotics, has been achieved. An English philosopher named Thomas Hobbes, in 1650, first proposed the notion that thinking is essentially a rule-based exercise. [1] Sometime later, in 1854, George Boole published *An Investigation of the Laws of Thought, on which are Founded the Mathematical Theories of Logic and Probabilities*. Others that followed continued to further the academic disciplines that supported AI and robotics. Eventually, in 1956, a small group of scientists met at Dartmouth College. At that meeting AI was born out of the discussion that ensued and subsequently became an identifiable discipline that we are now discussing.

Humans have always had a fascination with the physical basis of thought and behavior, and medical science has been concerned with the relationships between the mind and disease since the dawn of interest in these subjects as identifiable entities. These attractions have quite naturally led to the investigation of how the brain works and how the mind solves problems. The engineering correlates and paradigms of these pursuits are found in such current topical interest items as neural networks (analogous to brain performance), AI (analogous to human problem solving) and robotics (analogous to human mechanical movement) [2], [8]. One of the major contributions of these three disciplines has been the integration of developing knowledge about humans so that we can emulate ourselves.

Not surprisingly, medical libraries are replete with journals dealing with all of these subjects. The reason for bringing up this correlation of brain (reasoning) and behavior (functionality) is to point out that human inventions, such as communications networks, are benefiting and will benefit from advances in AI, neural networks, robotics, and the like. In addition, techniques of diagnosis are as valid for man-made systems as they are for the human body.

In-depth scientific investigations of correlations between mind and body have only been carried out during fairly recent times. The first proposal for mind and body integration occurred in 1947 with the definition of a new scientific endeavor called neuropsychology, fathered by Dr. Carl Pribram. And about 30 years ago, the idea of infusing intelligence into a machine was described by Alan Turing in a paper entitled "Computing Machinery and Intelligence." [10] This paper posed the question of whether machines can be designed to think. Problem solving is usually the stimulus for thinking. For his efforts, Turing subsequently became known as the father of AI.

The management of networks, whether they be computer, telecommunications, or a combination of the two, is concerned with the five CCITT functions defined for network management. However, probably the most fertile area for the effective use of expert systems in network management is in the area of expert maintenance systems. This technology provides the means to increase the accuracy of problem identification, and decrease the time expended in such efforts.

7.2 The Problem-Solving Issue

Humans are goal-oriented creatures. They have certain needs as well, such as the periodic satisfaction of hunger and early goals were directed toward the satisfaction of these needs. Humans pursue their goals and achieve them as best they can using their problem-solving behavior because there are usually obstacles to overcome. Other forms of life also possess these attributes although not to the same extent. So what is the difference between humans and other forms of life? The answer lies in the fact that, for the most part, humans can evaluate their progress, alter plans or develop new solutions, and, if necessary, adjust their goals to be more realistic. The ability to think in abstract terms does not hurt either. If we are to describe how a problem is solved, such a narrative might go as follows [8]:

(1)   Define the goals.

Possibly the goal is to diagnose the problem with a car or in a communications network. In particular, the goals may be to reduce the areas of ambiguity where possible problem causes may reside within the telecommunications network. Goals are usually based upon forms of conflict that involve abstract interests, such as personality needs. These goals may be long-term or short-term depending upon the individual's commitment to differing periods of time and urgencies of need.

(2) Define an approach to the solution.

This might entail observing the sounds and appearance of the car, using tools to probe and pick at certain places in the car, or using diagnostic equipment to determine the problem. In network management this might entail the use of some purchased network management system to test and diagnose certain features, or to monitor the performance of the system. Obviously, there is usually more than one approach that can be employed, although only one is pursued at a time.

(3) Define the criteria for success.

The criterion in this case is the isolation of the problem. These criteria can be expressed in rules that are linked so as to identify success or failure. In the telecommunications environment, this might be the specification of a 99 percent up-time criterion for certain equipment, or the availability of some function a certain percentage of the time.

(4) Collect data from the object and/or the environment at large.

This might include observations about the vehicle in various stages of rest or operation. Another way of viewing this assertion is to say that we must pursue the objective. The ongoing operation of the telecommunications network would allow data to be sampled periodically, from the equipment and tasks involved.

(5) Evaluate the results.

Results are determined from a comparison of collected facts and the rules for success. If the collected facts in a telecommunications environment show that the criteria are not being achieved, then some action would need to be taken.

(6) Be flexible.

If the facts do not yield a signal of success, then the process must be altered to collect more facts or to change the rules that lead to success or to the goal, i.e., sometimes we have to alter our expectations.

(7) Be prepared to develop new approaches.

If several attempts do not yield a measure of success or goal realization, then the approach may have to be reexamined to determine if a different attack should be taken to bring about success. If the network management system does not yield the required criteria for success, then it may have to be swapped for another system or substitute methods may have to be invoked.

These abilities, to be aware of insufficient facts, to realize the need for the reexamination of the approaches, and to comprehend the potential problem of mismatches between facts, goals, and errors, may be the chief differences between humans and animals. It is not clear, by the way, why humans always have to compare themselves to animals. Possibly, it is because of some sense of inferiority, but that is another story.

The preceding seven problem-solving steps point out the basic problem with network management systems, and that is these types of systems tend to be very difficult to define in detail. As a result, there is a trade-off that must be addressed, such that the users of the system have to decide whether they are content to infer diagnoses from macrolevel indications, or whether they are willing to spend the money to get at the microlevel data from which more definite conclusions can be drawn.

Another attribute attained by humans, thus far, has been their ability to conceptualize an approach or even several approaches to problem resolution. In other words, humans have been able to think in abstract about how the problem could be solved without previously testing that idea or hypothesis. Thus, once the approach has been developed, then it can be tested and reevaluated for validity or relevant truth. Hence, surrogate human capabilities, as expressed by certain AI software or robotic mechanisms can be conceptualized and designed without ever having seen such items in action.

The progress of AI and robotics can be measured to a great extent by the progress of automation. [8] The exercise of simulated thought is only possible using some form of computational device that can "weigh and sift" through all the bounded alternatives. [9] The process of exercising intelligence over the resolution of problems can be broken down into some generic steps, these being: (1) defining goals; (2) identifying alternatives; (3) collecting data, some of which may subsequently be discarded as being inappropriate or irrelevant; (4) reaching conclusions based upon the comparison of information with the goals leading to a refinement of the data collection process; and (5) comparing the conclusions with rules that constrain or bound the original goals. In cases where conclusions are not appropriate, the alternative may be adjusted and the data collection and ensuing comparisons reworked. This process is much like that used in the scientific method for experimentation.

AI has been proposed as a solution or part of the solution for just about every known problem. The many different initiatives that have been proposed include using AI as the work horse for financial decision making, manufacturing, and military operations. But the uses we are most concerned with here are the development and use of expert systems to mitigate such labor-intensive tasks as operations, maintenance, and administration support, as well as other more intellectual activities like trend analysis and prediction. The potential for financial impact in these areas is enormous, and some would argue that the long-term future for AI and robotics alone entirely justifies almost any expenditures for the results that will surely be attained. A more detailed evaluation will be provided later in this chapter. Nowhere in the progress of expert systems development has there been more promise than in the pursuit of expert maintenance systems.

Expert maintenance systems have also had an interesting history by themselves. A few years ago, General Electric's diesel locomotive division in Schenectady, N.Y. was faced with the impending retirement of its chief, and most experienced, maintenance expert. This gentleman had been with the company for some 30 plus years and there was little about these huge machines that he did not know, especially their operation, symptoms of failure, as well as the causes of such telltale symptoms. Faced with this problem, the company tried several approaches to "capture" this man's knowledge so that functional and operational continuity would not be lost to the maintenance operation. The interesting thing about this problem was that the best solution ended up being one in which several persons were assigned to interview the retiree with the express objective in mind of "picking his brain."

Thus, the interviewers were able to document his thought processes within this domain. He provided all the steps and circumstances surrounding the possible scenarios of locomotive maintenance and/or failure of which he was aware. This information was later incorporated into a rule-based computer system for others to use in performing their maintenance functions. This was the first instance of a real-life implementation of an expert maintenance system, and it worked. Today various maintenance "associates," as these intelligence programs are sometimes called, use workstations incorporating optical disks to provide rule-based, problem-oriented video tours of the specific maintenance task for the worker, while the system also displays maintenance procedures on the screen.

The interview and information extraction process, now part of what is known as knowledge engineering, followed a model that has been replicated over the years in many development efforts. The typical AI system has three main components, namely a rule base, such as that collected from the retiring maintenance expert, a continual set of facts that are incoming into the system from the environment, such as symptoms of locomotives in for diagnosis, and the inference engine that compares the facts with rules to reach conclusions about the subject. [7] These three components interact to evaluate real facts, or to collect an uncertain picture, to apply rules or heuristics to these facts, and to reach conclusions based upon the uncertainties involved. The rest of this chapter will explore the techniques and applications of expert maintenance systems, using the principles outlined here.

7.3 The Maintenance Problem

Most of the state-of-the-art equipment used in telecommunications has designed-in fault and operations monitoring capability. Unfortunately, many fault indications tend to be false alarms. And of those that are, in fact, valid many cannot be replicated during preliminary field testing. Other fault indications come from sources that are intermittent. Additionally, the projections for the maintenance force of the future is that it will decline in technical capability. And finally, of those system problems that are traced, the level of ambiguity in isolating the offending line replaceable component is many times much greater than one.

The message here is that the maintenance problem is large and growing. Current-day methods of quickly isolating and replacing failed units are lacking and tending to get worse, if a more intelligent approach is not instituted to perform the maintenance functions. In one military communications system, false alarms exceed the real alarms by an over 10 to 1 ratio. Of the real faults detected, one-fourth could not be replicated by field testing equipment. Of those that could be detected, 98 percent were detected by automated means, and the other 2 percent were detected by manual means. Of the units replaced in the field, half checked as being good at depot maintenance. This situation indicates that automated means of fault detection are not doing the job sufficiently to eliminate false alarms and to reduce the ambiguity level to offending units down to one. [2]

The two biggest maintenance problems are false alarms and unnecessary maintenance on units that are actually still functional. A survey of the U.S. and Canadian military shows that over 30 percent of units sent to maintenance for repairs tested good. We have already seen how false alarms affect the fault reporting rate. Expert maintenance systems can contribute significantly to the resolution of both of these problems.   The following sections will outline how this can be accomplished.

The question often arises as to how chronic maintenance problems are uncovered to begin with. The answer lies in the fact that, over time, records will convey the problems through reorders of bench stock or one-time orders, maintenance records will reflect the nature of repair actions, numbers of reorders will reflect reliability, etc. These issues can be collected in operational reports that are so important to the ongoing stability of the operation. Certain chronic problems associated with telecommunications network operations are difficult to isolate, and problems are usually identified when expected results are not generated from known conditions.

7.4 Requirements of Expert Maintenance Systems

The effective *expert maintenance system* (EMS) is a lot more than just a computer program. It is the result either of designed-in hardware sensors or interfaces, or of special-purpose hardware designed to interface with the existing system [4]. It is also the template for an even more involved enhanced maintenance training system, sometimes called the *computer-aided instruction* (CAI) system. Two of the key data collection ingredients for AI applications in maintenance are the *built-in test* (BIT) and *automated test equipment* (ATE), both of which are used to identify the problem so that the EMS can isolate the unit's fault.

The built-in test unit is, as the name implies, built into the operational unit itself. Thus, the BIT as well as the operational unit are both designed with the other in mind. The ATE, on the other hand, is usually separate equipment attached to the operational unit for purposes of testing when testing is warranted. The use of this equipment, however, does not guarantee success in isolating the problem at hand. The expert system helps to refine the findings of the traditional BIT and ATE systems. Such systems are basically used to detect operational values. The expert system, on the other hand, has requirements that are outlined below. [5] CAI is a combination of technology and procedures that allow the maintenance person to understand and diagnose the problem within some criteria. These criteria are usually time and efficiency. Efficiency, in turn, can be defined as employing the minimum amount of replacement inventory as necessary. CAI, therefore, requires an in-depth understanding of the system operation, its component parts, how these parts interact, and how the maintenance person wants to approach the problems at hand.

7.4.1   Knowledge Base System

Database, or knowledge base design for the support of expert systems inferencing, has been a legitimate pursuit in computer science for many years. [5] The major difference between a database and a knowledge base is that in a database the retrieved object is a fact, whereas, in a knowledge base the retrieved object is a deduction. In addition, the knowledge base must understand queries and respond in natural language fashion.

7.4.2   Theorem Proving

The expert system must be capable of proving an assertion, or theorem, e.g., that a particular fault x occurs at location y. This is somewhat akin in an opposite sense to the null hypothesis where the assertion is that there is no difference between a particular event and its consequences. Theorem-proving techniques attempt to establish procedures for selecting among several rules to apply to the problem and to establish problems leading to the solution. A theorem usually has a certain level of uncertainty attached to it, so that the proof, in turn, has a certain level of probability attached.

7.4.3   Automatic Programming

All compilers are really *automatic programming* tools, because they convert certain syntax into executable object code. Such code is the list of instructions that can be directly read by the computer and executed as ordered. Automatic programming takes this concept another step further. In automatic programming, high-level descriptions of need are themselves processed into executable object code. These high-level descriptions may be loosely structured language that is processed through a natural language processor, then interpreted, and finally subjected to an object code generator.

7.4.4   Solutions to Combinational and Scheduling Problems

The traveling salesperson problem has been one that has received considerable mathematical interest for many years. It is one in which the issue is how to schedule the rounds of a traveling salesperson so as to maximize his or her customer exposure over some period of time, or said another way, to minimize his or her travel time between customers.

Solutions to these and related problems generate an explosion of possibilities. Typical solutions require an increasingly large time frame in which to resolve the near-optimum possibilities sought. AI approaches operate so as to constrain these solution possibilities by using domain information. Such domain-specific information might be appointments or other needs that would negate some of the possibilities.

7.4.5   Machine Perception

There are two basic forms of machine perception, one being vision and the other being language or speech. The understanding of human-spoken language by a machine is an example of speech perception, and vehicular terrain-following perception is an example of machine vision. In the field of telecommunications both types are of equal importance. Mapping and exploring, geopositioning, and object location are just a few of the possibilities for machine vision. With regard to speech perception, the receipt, decoding and interpretation of language is of obvious interest to telecommunications. Many speech perception systems have already been developed, many by the toy industry. Such systems will find their way into more widespread use in the near future.

7.4.6   Expert Consulting Systems

Such systems are oriented toward solving problems and many times can be an adjunct to an existing on-line expert maintenance system. These systems interact with the human expert to receive rules, or may carry on a dialog with the human operator to collect ancillary data, or might test specific hypotheses requested by the operator.

7.4.7   Robotics

Robotics is an exploding technology in the product and service delivery arena, and will impact telecommunications to a great extent in the next few years. Potential areas of usage are maintenance and repair, operations monitoring, and security.

7.4.8   Natural Language Processing

This is such an important topic from a human interface point of view that it was covered in the previous chapter. The basic objective for this line of interest is to make the machine capable of understanding unstructured inputs from the human operator, whether they be spoken or typed.

7.5 Real-World EMS Issues

The application of expert systems capabilities to maintenance carries some real-world issues that must be addressed. [6] The major ones are discussed here. Expert systems technology begins and ends with a thorough understanding and application of the techniques involved with the technical elements. There are two ways of viewing AI; one is to use it as a means for solving technology problems, and the other for solving human performance problems. This section will address both. [3]

Basic Skills

The skills trend for the foreseeable future is not good, because there appears to be a diminution of worker skills in various categories over the next few years. Those qualified for technical training will decrease in number and will have entry level skills such that emphasis upon computer assistance is almost mandatory. The skills required include the abilities to collect facts, pursue logic paths, and interpret intermediate results. These skills are exactly what the well-thought-out expert system addresses. Thus, declining human skills can be bolstered by the expert system approach.

Simulation

The use of simulators is a long and time-honored method of training people for technically oriented tasks. The use of simulators is an approach to raising the qualification level of the humans involved, not the injection of expert systems technology into the process. Current-day simulators have reached new heights of realism and variability for operations or maintenance scenarios. One need only experience or even see the simulators used for aircraft, marine, or submarine training to appreciate the breadth and depth of training through simulators. The latest simulators using virtual reality have essentially integrated reality with technology-created surroundings and circumstances to produce physical, mental and emotional surrogates for reality.

Computer-Aided Instruction (CAI)

Learning is a process best enforced by a variety of methods, some of which are programmed instructional training, simulation, hands-on practice, classroom teaching, visualization through such media as video programs, and CAI, to name a few. In this approach the instructor's expertise is captured and replayed to the student to be tested against the expert. There are varying degrees of effectiveness gained from such means, because the realism is more limited than the simulator, however less expensive.

Development of Instructional Material

AI systems can be used to screen, and evaluate material collected as a result of scheduled maintenance as well as unscheduled outages, for the purpose of developing a framework of required training curricula. This framework can also be expanded to refine classroom training plans and topics of ancillary importance to the overall training program. Additional expansions of the basic paradigm are for the development of learning models by an AI system that will match up with skills and categories of technical positions to be used.

Maintainability Design

There are many people who believe that the average, current-day, high-tech appliances, whether they be commercial or military, are too complex for even the above-average technician to maintain. Considerable work has gone into efforts to develop physical and/or functional schema to overcome this perceived problem. For the most part, efforts at structured software design for the purposes of enhancing maintainability, as well as rigid hardware design constraints, have failed to achieve the increased maintainability desired. However, several approaches to the problem are being pursued, among which is the use of cognitive models. In the meantime, continued work will be performed to understand better, and to impact maintainability vis-a-vis the design approach.

Cognitive Models of Maintenance

The standard tree-structured approach to troubleshooting hardware/software functional problems is slowly being replaced by what is known as cognitive modeling of the device to be examined. Cognitive modeling is a method of reducing the functional problem by performing the most important and useful tests first, followed by other more specific or more esoteric tests depending upon the results obtained during the initial testing period. The whole idea is based upon the way that humans would test the device if they had the requisite knowledge to interpret the results, but were pressured in terms of time or resources. Essentially, the technician has a collection of tests and a collection of replacement units available. The trick is for the technician to be able to evaluate a particular test or a particular replacement action in terms of solving the problem. As alluded to above, the process is iterative, in that it performs the most useful tests first, followed by less likely tests until the issue(s) is/are resolved. Obviously, there may be multiple failures that complicate the problem and may present an expanded number of possibilities.

Automated Documentation

The documentation of technical design or processes is a problem area, because of its close association with human weaknesses for missed detail, as well as erroneous inclusions. A very promising attempt to resolve such human errors is to provide the technical person with a means of automating authoring capabilities. Systems that provide an author's workbench, or author's associate, where computer-determined language is inserted into the documentation stream, are a good start, but much work remains to be done to make such efforts truly effective. Deeper-level A.I. algorithms will enhance and solidify the codification of technical documents for the technical reader.

AI Applied to Real-World Maintenance

There are couple of issues to be addressed when tackling the real-world problems of applying AI to maintenance. First, the AI program must be robust; i.e., it must withstand breakdown or getting lost when confronted with symptoms on the periphery of its domain. The line between one domain and another is usually gray, or ill-defined. Therefore, the domain specific program that pursues problem specifics has to be protected against unknown or extra-domain symptoms. This is a significant challenge for the knowledge engineer and programmer. Second, the objectives of the expert system must be worked into the system design and implementation. Examples of valid objectives are safety, repair, and/or diagnosis. The improper prioritization of these options can be serious in more ways than one.

7.6 Approaches to Maintenance Troubleshooting

Troubleshooting device failures accounts for about half of the maintenance routines involved in computerized maintenance. There are about six different ways of implementing a troubleshooting maintenance system. Some of these are described below. [4]

7.6.1   The Self-Contained Computer Program

The more common approach, used currently, is to generate elaborate computer programs that actually contain the system's characteristics and indicators, i.e., its knowledge, in the code itself. Aside from the problem of lacking portability from this particular application to another circuit, line replaceable unit, or system with similar characteristics, this approach also lacks in other regards. It is difficult to categorize the faults and/or conditions being attacked. Second, the code really doesn't provide the visibility into whether the faults addressed were in fact operated upon. Finally, the faults addressed really don't provide insights into whether they were the important faults to begin with. Therefore, the troubleshooting system using this brute force approach lacks robustness, because it is non-transferable and bypasses the question of inference. Inference, in this regard, is concerned with the questions of following the most important paths first and covering measurements that yield the most information second so that subsequent measurements and path selections can proceed as efficiently as possible.

7.6.2   The Rule-Based Inference Engine

The next, more sophisticated, form of troubleshooting is called empirical association. These empirical association rules are derived from experts, or at least people that have long-term, in-depth acquaintance with the system in question. Interviews with such people show that conversations surrounding the issues of maintenance invariably come back to comments like, "*If* so-and-so occurs, *then* such-and-such is the situation, *else* what-you-call-it will occur." These *if-then-else* statements form the basis of the empirical associations so important to inferencing.

Unfortunately, empirical associations also have their own set of drawbacks. The collection of rules must be organized. First, all rules that apply to the remaining set of candidate problems must be pulled from the total ensemble. Next, the candidate sets are further divided up based upon additional rules pulled from the total set that provide maximum information about these candidates. The increasing granularity of the candidate space must at each step be evaluated against the costs involved for practical situations. Other problems include the large number of rules required for even relatively simple systems, the problems of adding or subtracting rules, guarding against adding rules that may contradict rules already in the rule base, and assuring that all potential faults are accounted for by the rules employed.

There are, however, certain advantages to the empirical associations approach. The information concerning actions and consequences is now part of the rule base, instead of being implicit in the troubleshooting program itself. Also, the rule base is easier to modify than the previous computer program approach. And finally, the efficiency is improved, because the information content improves faster as new measurements are made, due to the fact that now, new measurements are dependent upon previous measurements as opposed to some preset path or methodology where measurements are incidental events.

7.6.3   Causal Expert System

If the rules in the rule base are organized so that they can be collapsed as needs require to define a path based upon the causes behind functionally abnormal outputs, then this method defines a causal expert system. In causal systems, paths are chosen and paths taken on the basis of structure and the attendant causes for the information needs implied.

7.6.4   Behavioral Knowledge About Devices

The idea here is to look at the behavior, or the outputs, of the functional interfaces in the device. Any abnormal or unexpected outputs trigger an examination of intermediate outputs until the problem is uncovered. Typically, these abnormal or unexpected outputs cause one of the system components to come under scrutiny. Such components are referred to as a *unit under test* (UUT). An example of such an approach is illustrated in Figure 7-1. [1]

If the output at z is determined to be faulty, the intermediate outputs at $f(x)$ and $g(y)$ are examined to determine the real faulty component. Thus, the original faulty indication provides a considerable amount of information with which to judge the next action. The original choices for the fault(s) are components f, g, and h. Even if all three components have to be examined, the system has been able to isolate the problem to these three components as opposed to examining all components in the domain.

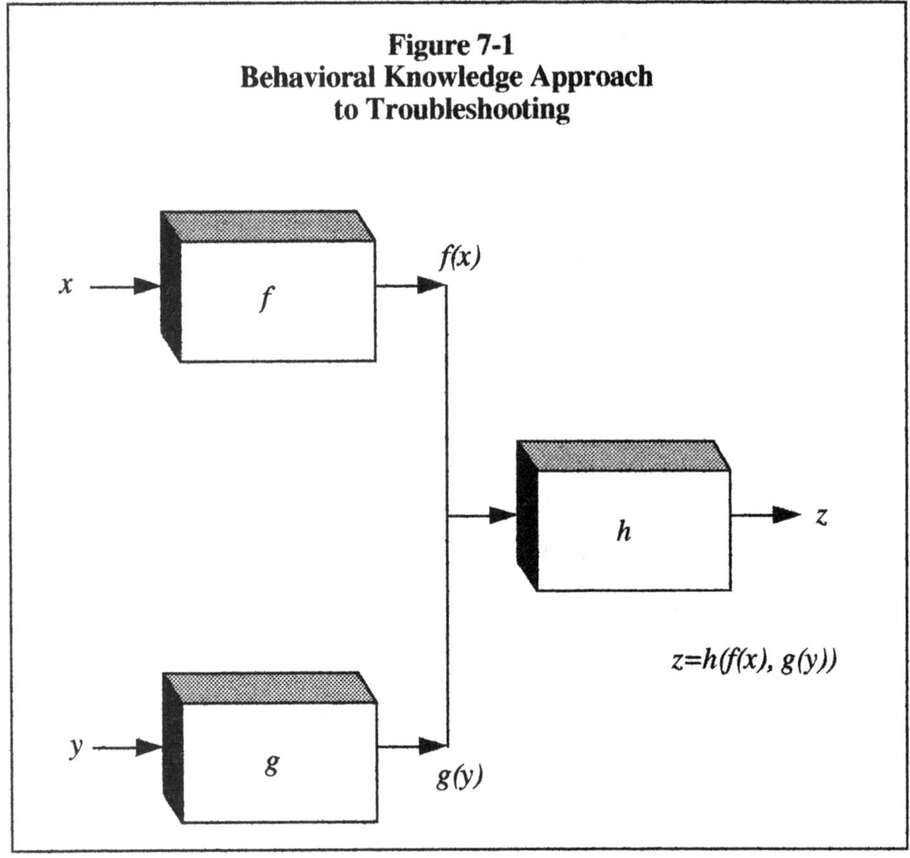

Unfortunately, the solutions are never quite this easy to effect. First of all, there may be two or more units at fault, or the level of ambiguity may not be capable of being resolved down to a single unit, the testing of specific units may not be possible without affecting the outputs of other units, and so on. The ideal knowledge set about the behavior of a UUT has to start with the design phase of the system, its subsystems, and specific components. At this point, the signal and process flows, input and output paths, and conditional responses can be designed into the system in such a way as to minimize ambiguity and maximize problem isolation potential.

The difference between descriptive and behavioral knowledge underscores the various approaches that can be used in the pursuit for diagnostic truth. The need to deal with the behavioral approach arises from the fact that problems are seldom presented through a description of the system or UUT, either physical or functional. Failures result from behavioral aberrations of the unit. Thus, behavioral information is essentially an interface issue that comes as a result of translations from the physical and functional manifestations of the problem at hand.

7.6.5 In-Depth Fault Mode Analysis

In this approach the objective is clearly as stated, namely that fault modes are examined as opposed to output expectations. By way of example, consider the situation depicted in Figure 7-2. Here, the fault diagram portrays a series of inputs and outputs all of which depend upon the input or output of element A. Element C is independent, and therefore, does not affect the results of the output at E. The outputs of B and D impact E, but therefore, a fault at E can only originate at A, B, D, or E. The tabulation below explores the principal issues of fault finding.

	Input	Output
A	bad	error @ A,B,C,D,E
B	bad	error @ A,B,C,D,E
C	bad	error @ A,B,C,D,E
D	bad	error @ A,B,C,D,E
E	bad	error @ A,B,D

In this case, the fact that, assuming there is only one fault, the input to E is in error and inputs are normal elsewhere, the problem can only reside at B, D, or A. On the other hand, if an error is detected anywhere in the system, the best place to start is module A in order to maximize efficiency of search and to minimize time involved in the analysis.

7.6.6 Topological Reference

The last approach to troubleshooting involves a schematic diagram that is read into the system, its symbols are interpreted, its interconnections are defined, and its operation is inferred. The existence of trouble is marked on the screen of the system depicting the schematic, the topology is examined, and inferences about the system's failure are made based upon the graphical picture of the device that was captured. The vision, therefore, is to "read in" a schematic and to diagnose faults based upon indicated errors within the domain of the graphical boundaries of that schematic.

7.7 Maintenance Systems Models

There are several ways of representing knowledge within the confines of expert maintenance systems. The models that represent such approaches are discussed briefly below. These methods are not to be confused with the troubleshooting approaches discussed above, because those methods are about how the designer addresses the problem as opposed to how the problem is actually solved in the program, as indicated here.

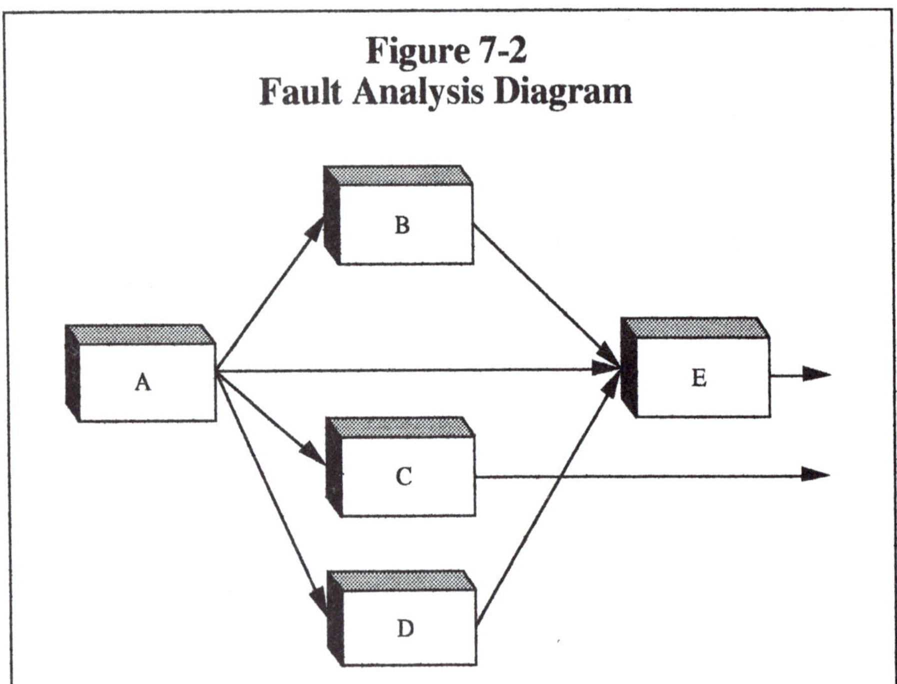

Figure 7-2
Fault Analysis Diagram

7.7.1   Information Model

Here the symptoms are collected, and a group of components are considered based upon their relationships with the symptoms uncovered. The probabilities of abnormality are based upon the cumulative probability or complexity of the error situation possible. The component(s) with the highest risk of error are then tested for actual error conditions. Figure 7-3 illustrates this stair step process. After the final step, the process is evaluated and the necessary steps repeated. The complexity issue is introduced to point out that there are situations in which the problem resolution takes more time than others. Many believe that the reason for such situations is because of the complexity of the components involved or their interactions. This model allows us to correlate maintenance actions with the times involved in actions of varying complexities.

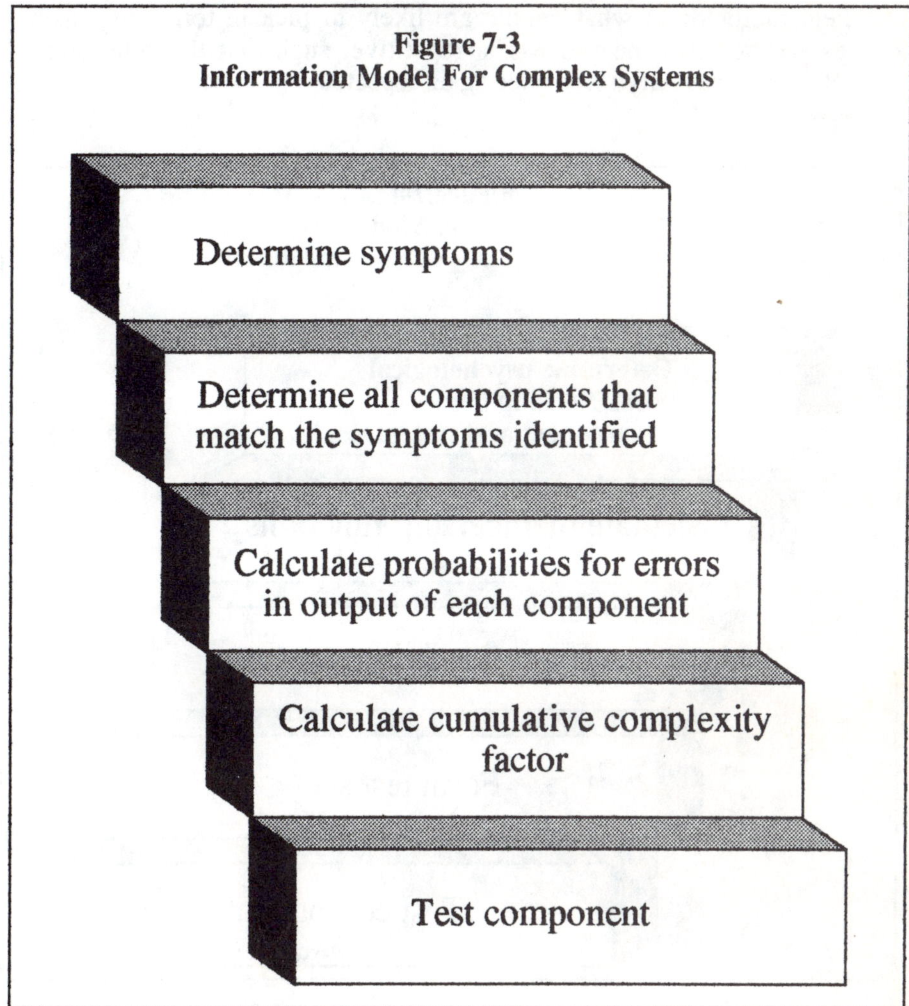

Figure 7-3
Information Model For Complex Systems

7.7.2   Fuzzy Set Model

The information complexity model shows that the complexity of a problem is as much related to the problem solver as it is the system. Another level of granularity is concerned with the nature of the actions necessary to be performed in resolving a problem. This approach is about the feasible set of actions that might be likely and the choices to be made within that feasible set of possible actions. People are not very good at determining feasible sets of actions. They tend to derive a "fuzzy feasible set." People more often come up with heuristics, or rules of thumb, that frame the issues for them. The fuzzy feasible set is indicated in Figure 7-4, and is an excellent predictor of what people are likely to pick in terms of feasible actions. Here also, the process is repetitive, such that the final step is evaluated and necessary steps once again repeated.

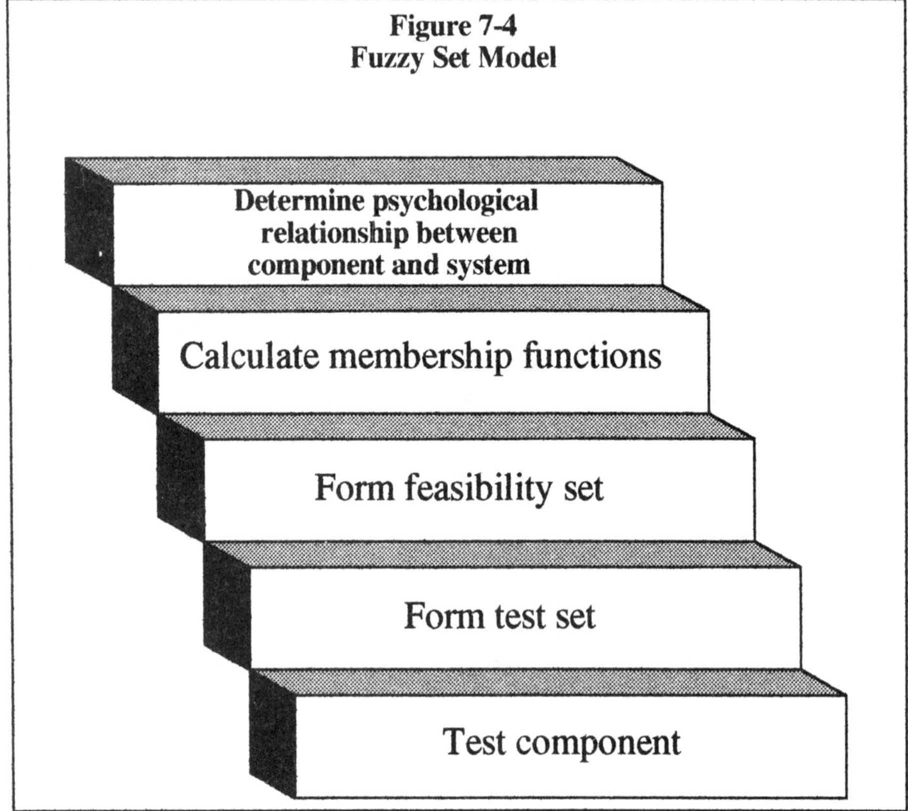

7.7.3   Rule-Based Model

The rule-based model is one in which people approach their problem with a set of heuristics or rules of thumb. Action prediction can be made with a accuracy of about 90 percent. The rule-based approach to problem solving tends to indicate that people know the rules to be applied, but lack the timing or knowledge of when to apply such rules. The rule-based approach is summarized in Figure 7-5. The various strong and weak points of this model have already been explored; however, it remains the most used approach to expert systems implementation yet today. The rule-based system has two major filters used in the resolution of problems, these being the fact base, or comparator, that is used to detect abnormal conditions, and the other being the rule base that evaluates system parameters as a means of isolating the offending element. In practical situations concerning the data collection about the environment or system operation, these data can be determined either from automatic data collection methods, or from manual inputs about the system operation.

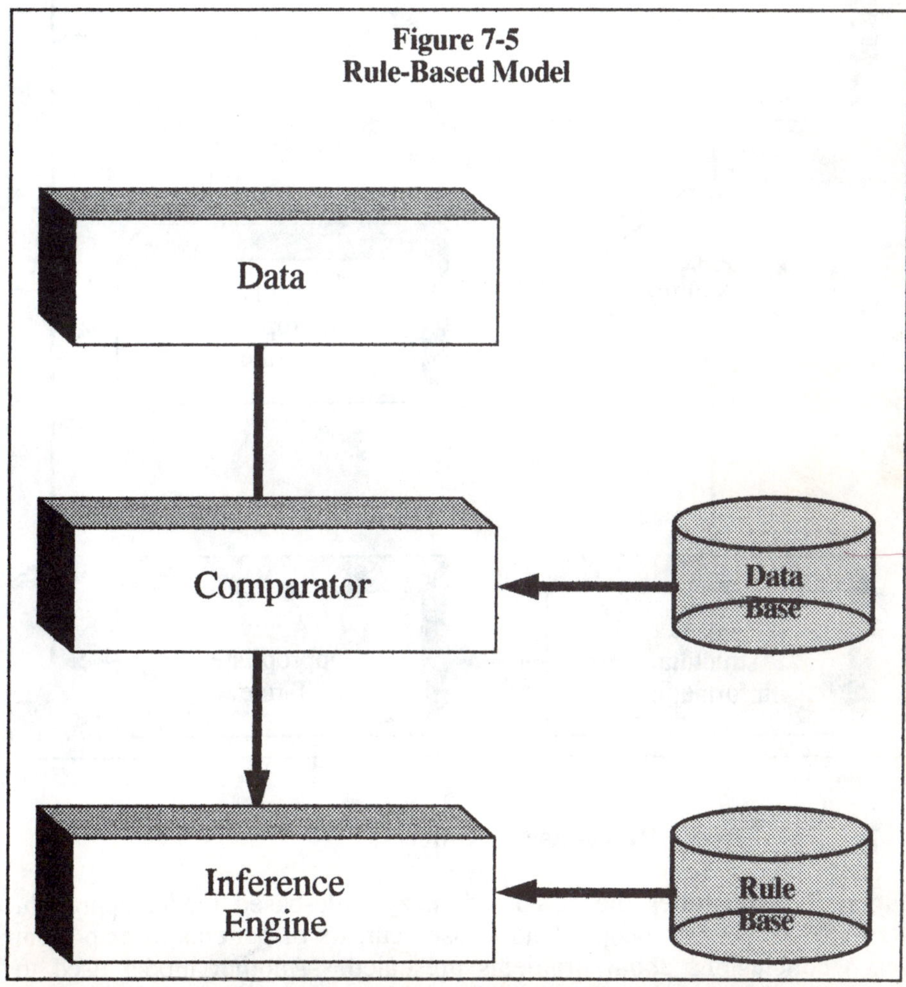

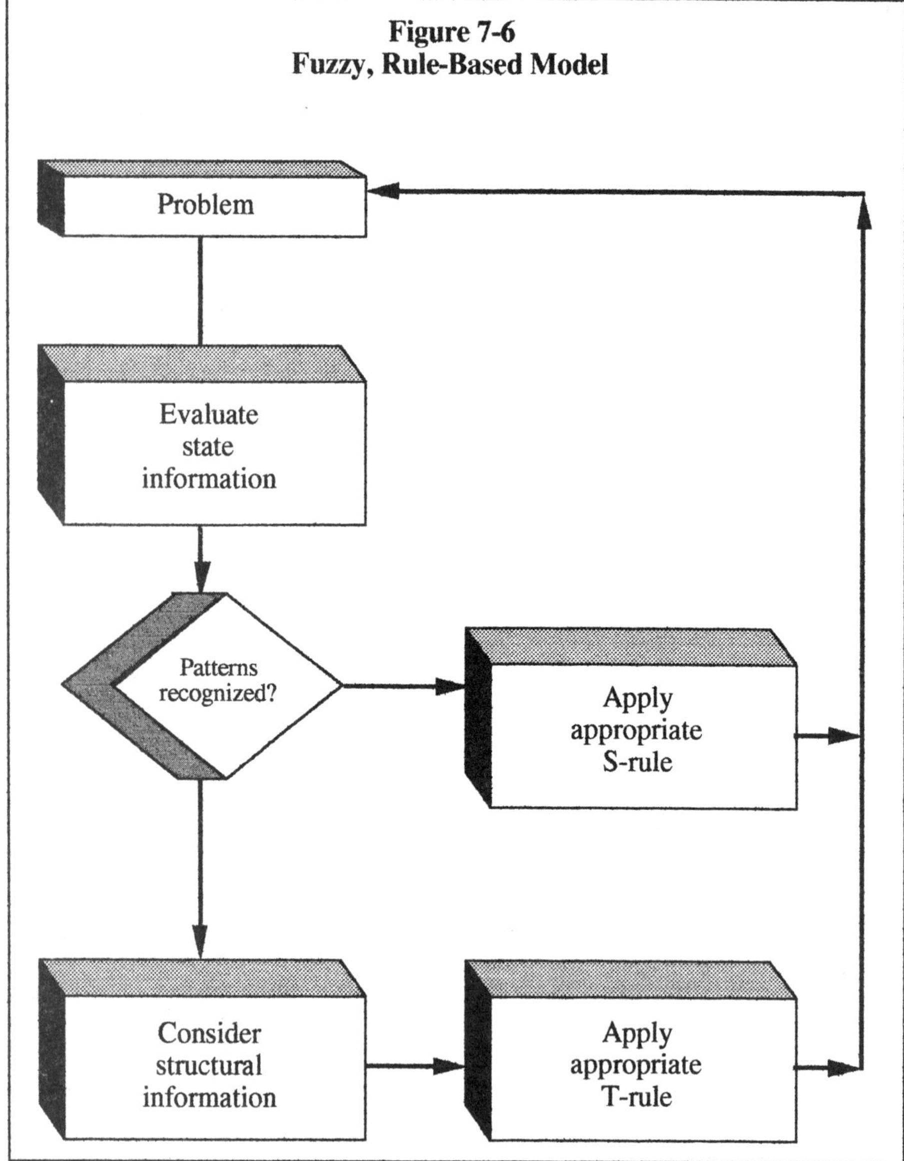

7.7.4   Fuzzy, Rule-Based Model

Figure 7-6 illustrates the combined fuzzy, rule-based model approach. Research shows that people tend to use contexts or patterns to help them reach conclusions about problems presented. Another model used to approach problem solving is one that combines two of the types previously discussed, namely the rule-based approach, and the fuzzy set approach. Using this model, the problem solver considers the system at some point usually called some information state. Various patterns familiar to the solver are examined, and if any pattern(s) emerge, an S-rule is applied. The S-rule is a symptomatic rule that means that a symptom is simply applied, and mapped to a solution. If no familiar pattern arises, the T-rule is applied. The T-rule is concerned with structural approaches, although people are not as adept at applying structural rules as they are state-oriented rules. T-rules examine the physical aspects of the system as opposed to the S-rules that examine the functional aspects of the system.

Another approach to the problem-solving dilemma is that associated with how people pick their rules for getting at solutions. Since there are different categories of symptoms and resulting rules to be applied, we must sometimes look at rules in different categories of potential problems. In this approach, the maintenance person evaluates the state of the problem information available, and chooses the avenue of picking a state-oriented response as circumstances dictate. Next, he or she will consider whether structural information is to be evaluated. If such is warranted, a structural response will be utilized. Figure 7-7 shows the model for a multilevel, rule-based approach.

There are three levels of problem resolution, these being problem recognition and classification, resolution planning, and execution and monitoring. Table 7-1 shows how these levels relate to the decisions and responses involved.

7.8 Knowledge Representation Issues

Some of the principal issues concerned with knowledge acquisition and utilization are covered in the sections below. The maintenance problem, as it is tackled through the use of expert systems technology, requires close attention to the means by which the solutions can be delivered. These means come principally by way of the representation of the knowledge supplied. This is why the topics outlined below are so important to the architectures of which they are a part.

7.8.1　Architectural Extremes to Be Considered

In any technology, architectural program, or system, there is always the struggle for the optimum approach to supporting the search for the problem solution. There are many such approaches, embedded in the

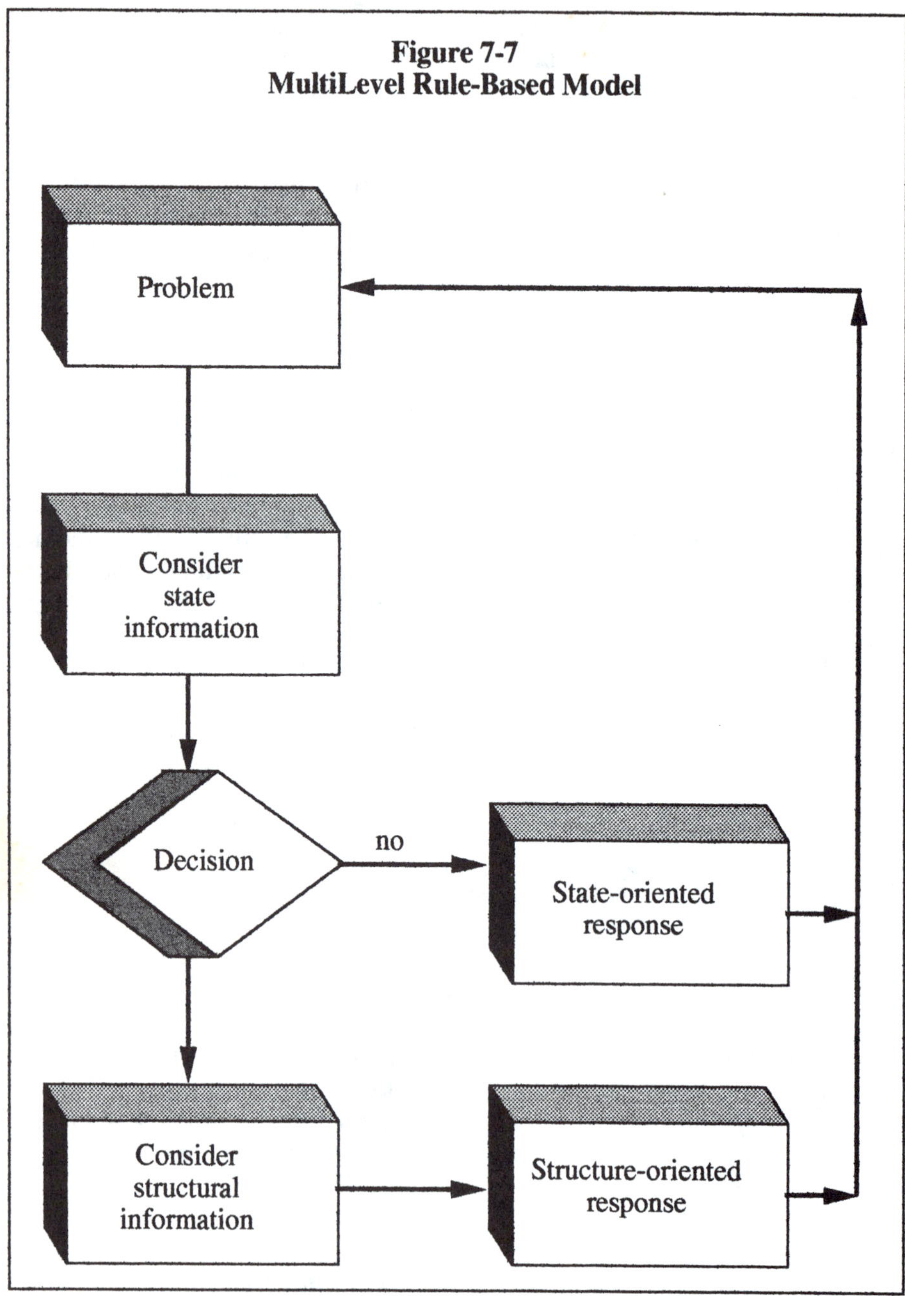

Figure 7-7
MultiLevel Rule-Based Model

application, all of which can be thought of as architectural embellishments that must be traded off against the specific needs of the system under development. These embellishments represent a range of possibilities as the designer begins and follows through with the system solution, and are highlighted below. The capture, representation, and use of knowledge occurs within the confines of these parameters.

Table 7-1
Decisions and Responses for Three Levels of Problem Solving

Level	Decision	State-oriented response	Structure-oriented response
Recognition and classification	Frame availability	Invoke frame	Use analogy and/or basic principles
Planning	Script availability	Invoke script	Formulate plan
Execution and monitoring	Pattern availability	Apply appropriate S-rule	Apply appropriate T-rule

---

(1) Search Space

The range of allocated search space can extend from a small space with its inherent increased data reliability and associated knowledge to an exhaustive search space where there tends to be a monotonic, single line of reasoning. The larger the search space, the more restricted and specific must be the search for data items related to the issues at hand.

(2) Knowledge Specificity

Here, the knowledge derived can range from totally unreliable to knowledge that is corroborated by combining evidence from multiple sources, using probability models to define the uncertainty involved, the use of fuzzy models that incorporate knowledge gaps, and even exact models of the system so as to identify what is appropriate from what is not.

(3)   System Transitioning Conditions

Transitions in a system can occur basically from one of two conditions, one being the transition of time, in whatever increments are appropriate, and the other being the transition of events or circumstances that alter system expectations.

(4)   Search Methodologies

Search methods range from random accesses associated with large search spaces to hierarchical searching, to relational searches. Random, large space searches imply a large ensemble of data items that can be accessed by direct addressing, as opposed to some indirect method needed due to confined or restricted space availability. Hierarchical searching is accomplished through a set procedure working through a specific, physical architecture. Relational searches occur as a result of predefined associations within the database that yield category accumulations of attributes when the search is initiated, calling for such attributes to be collected.

(5)   Partial Solution Evaluation

On the one hand, there are applications that provide constant evaluation during the solution phase. The steps in the typical solution process provide points at which not only progress, but also proximity to the solution can be judged. On the other hand, abstractions of the fixed-step paradigm may not provide intermediate opportunities for solution evaluations.

(6)   Real versus Abstract Search Spaces

The domain may be real and definable. In such a case, fixed sequence searches may be allowable, or may not, depending upon the size of the domain, the size of the search algorithm, or its complexity. If the search space is abstract itself, or virtual, search sequences may be fixed or variable, and of any length.

(7)   Layered Problem Solving

Some applications are designed to force the interaction between subproblems of various categories. This is especially important in situations where the solution is somewhat iterative. In other applications, there are collected or inherent constraints that prevent interactions between subproblems. Additionally, there may be circumstances where subproblems are disjoint within the same application.

(8) Solution Paths

Certain maintenance problems can only be initiated or contrived using pure guessing within the confines of the domain defined. Other situations allow plausible reasoning where educated guesses become the starting points for the solution. The ideal situations are those where there are defined dependencies that associate specific events and or structural realities to yield a directed approach to the solution of the issue at hand.

(9) Single versus Multiple Modeling

This is a question of whether the application should be programmed so as to accommodate only a single line of reasoning, as opposed to multiple lines of reasoning. The idea behind multiple lines of reasoning is that the solution may be derived sooner than later, and the solution may actually depend upon simultaneous paths of reasoning being pursued.

(10) Aggregation of Models

The application may easily accommodate diverse algorithms, or solution specialties; it may tolerate only one type or specific heterogeneous model of problem solution.

(11) Data and Knowledge Base

The knowledge base may be too restrictive for the application required, or the data structures may be too inefficient. The biggest problem likely to face the designer is that of which database to use for the particular application at hand. The data/knowledge base should have the flexibility needed, else the realization of the application required becomes difficult if not impossible.

7.8.2   Knowledge Repositories

The knowledge for maintenance systems development and integration into operational systems follows several paths. The basic first-order system for formal logic is that of predicate calculus. It is a method of assigning relationships between variables, and quantification of those variables. Such logic connections include negation, disjunction, conjunction, implication, and equivalence. For some traditional expert systems, production systems entailing if-then-else rules are incorporated into the operational facility to support the maintenance problem solver. Another repository for knowledge is the semantic net. It is basically a pictorial method of denoting the relationships between objects that are indicated as nodes, and their relationships, indicated as directed line segments. There are in turn variations on the semantic approach, some of which are: (1) equivalence semantics, where meaning is conveyed by similarities; (2) procedural semantics, where meaning is imparted by procedures; and (3) descriptive semantics, where meaning is derived from a dictionary or lexicon. Frames are essentially semantic nets except that procedures are added to the semantic nets in order to define the frame. Finally, we come to procedures that provide the steps for the processes concerned and are part of other knowledge representation methods.

7.8.3   Knowledge Acquisition

The knowledge acquisition process consists of several stages that are linked together to make a cohesive whole. The steps involved include identification, conceptualization, formalization, implementation, and testing. Figure 7-8 presents the process in its organized form as a series of tasks. The identification task is the point where the the problem characteristics are specified. The conceptualization step provides the concepts that represent the knowledge required. The formalization stage is that point at which the structural design needed to organize the knowledge is created. The implementation task is where the rules are formulated to utilize the facts available. Finally, the testing phase is the place where the rules that have organized the knowledge are validated.

7.9 Integration of Concepts

At this point it is appropriate to put together the various techniques and components of the typical expert maintenance requirement into a cohesive, integrated whole.

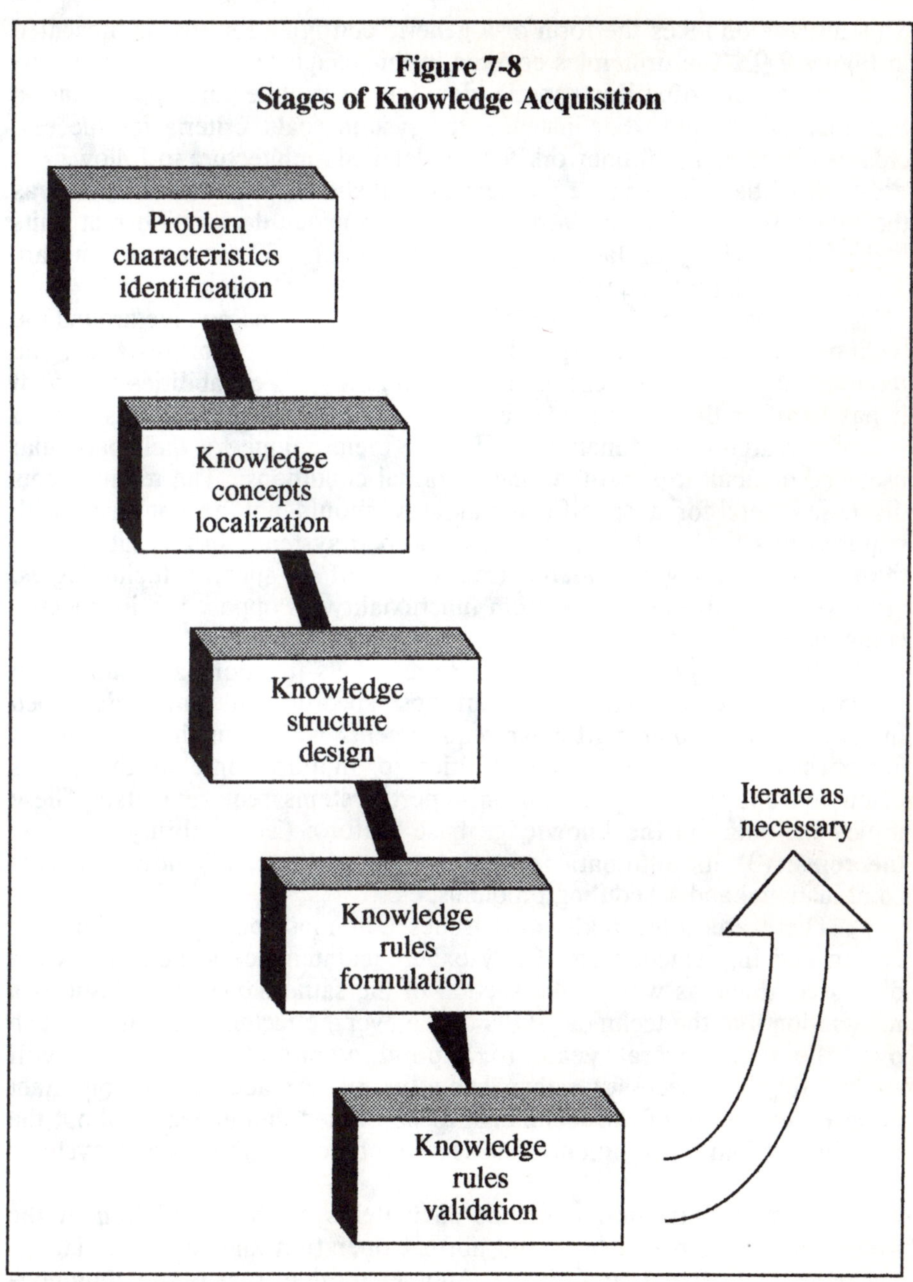

Figure 7-8
Stages of Knowledge Acquisition

7.9.1 Generic Fault Isolation System

This integration takes the form of a generic configuration that is presented in Figure 7-9. The principles covered in this chapter can be applied to the problem of converting this generic view into an architecture appropriate to the situation at hand. For instance, the system goals, criteria for success, etc., can become the framework for the detailed architecture to follow.

Next, the maintenance problem overall should be considered, such as the major issues of false alarms and the erroneous declaration that units have failed when, in fact, they are functional. These conditions are referred to as type I and type II errors in hypothesis testing.

The requirements for expert maintenance systems revolve around the technologies that these capabilities have to offer. For instance, the requirements for using a car revolve around the car's capabilities and what it has to offer the person on foot. Expert maintenance systems have a series of features and functions that make them unique for their particular uses and difficult to substitute under normal conditions. The requirements discussed here for a specific technology should not be confused with requirements that might be derived for a total system. In this latter case, the system is being considered irrespective of its specific technologies, because the focus is on the system functionality, as opposed to its specific components.

Once the system is defined in terms of its functions and subsystem components, then the requirements of those specific units can be described in terms of their own particular requirements which, in the case of the expert system, include its capabilities to support some of the topics discussed above in the section on expert systems requirements. These topics include: (1) the knowledge base feature; (2) its ability to prove theorems; (3) its automatic programming facilities; (4) how it solves combinational and scheduling problems, etc.

There are some real-world issues that must be included into the design and implementation of any expert maintenance system that were discussed above as well in the section of the same name. One issue is a recognition that the technical skills of the average technician will diminish over the next several years to a point where expert systems will conceivably be a necessity rather than a luxury. An additionally important issue is the ability of the technician to be trained and educated about the workings of the equipment that is the object of the expert systems operation.

These training tools include such items as the simulation of the system in order to determine its normal operation and to appreciate its behavior under failed conditions. Another training tool is the computer-aided instruction facility that is imparted to the student through a series of programmed instruction exercises coupled with scenarios to test the

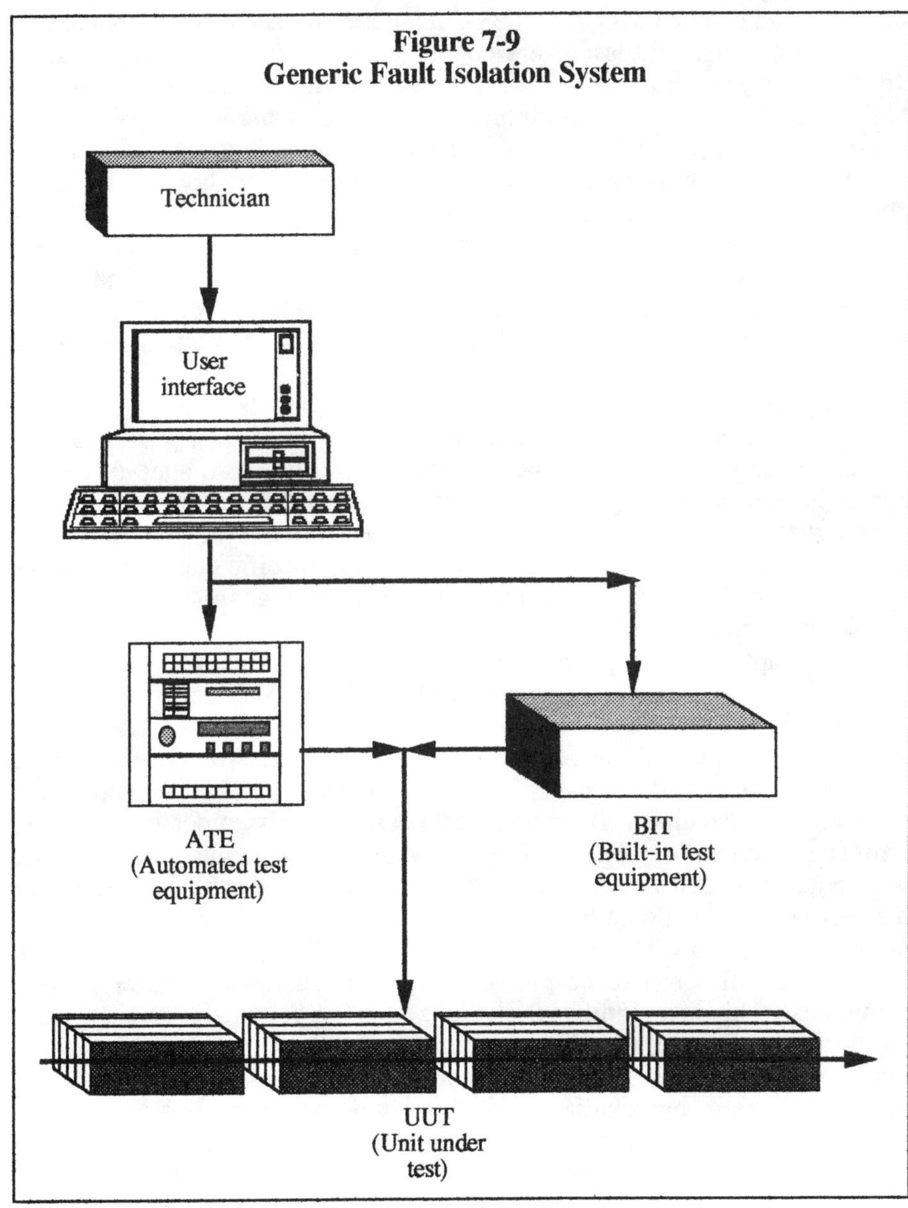

Figure 7-9
Generic Fault Isolation System

student's absorption of the material covered. Finally, a training tool that actually generates instructional material may be of great interest, so that the needs of the specific individual can be addressed by the computer training system. In such cases, the system isolates the student's weaknesses, and generates material customized for such an individual so that particular shortcomings can be covered while skipping the student's strong points.

Other real-world issues are concerned more with the design phases of the system than its instructional or maintenance problems. Maintainability design is a discipline where the designer is trying to design the system for ease of maintenance in parallel with its functional objectives. This process gets heavily involved with the subject of quality, but for these purposes we will simply think about designing to satisfy the real requirements, which is a significant goal for a quality product, along with minimizing the system's complexity, and grouping components into functional and accessible *line replaceable units* (LRUs). Cognitive models are also of importance as real-world issues concerned with the maintenance problem.

The old approach to troubleshooting, where an elaborate tree unfolds as the technician goes through one branch after another in pursuit of the failures point(s), is cumbersome to deal with. The more state-of-the-art approach is to apply cognitive rules or heuristics that cut corners for the expert that is familiar with the peculiarities of the system. Here, the system's history, uniqueness of symptomotology, specific conditions output by the system all lead to sets of diagnostic reactions that are based upon this cognitive approach.

To capture the importance of the final three topics to be addressed here, some repeat of what was stated above in the section on real-world issues is warranted. These three final topics are automated documentation, system robustness, and the design of the system itself. The documentation of technical design or processes is also a real-world problem area, because of its close association with human weaknesses for missed detail, as well as erroneous inclusions. The AI program must also be robust, i.e., it must withstand breakdown or getting lost when confronted with symptoms on the periphery of its domain. The line between one domain and another is usually ill-defined. Therefore, the domain-specific program that pursues problem specifics has to be protected against unknown or extra-domain symptoms. This is a significant challenge for the knowledge engineer and programmer. Finally, the objectives of the expert system must be worked into the system design and implementation. Examples of valid objectives are safety, repair, and/or diagnosis. The improper prioritization of these options can be serious in more ways than one.

7.9.2   Generic Architecture

Once the various issues surrounding the parameterization and design of the expert maintenance system have been dealt with, the generic architecture can be unfolded. An example of one such architecture is illustrated in Figure 7-10.

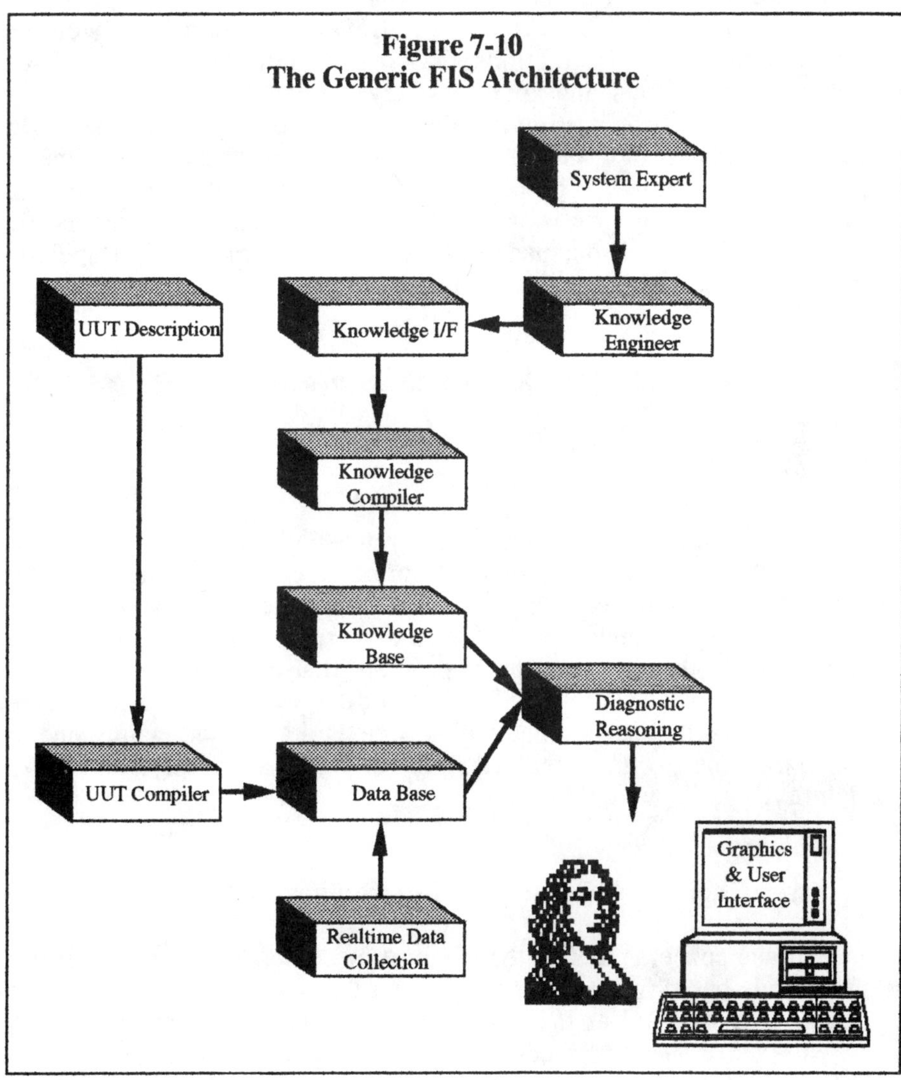

The key elements of this system, aside from the user and expert, are the knowledge acquisition interface, the knowledge compiler, and the diagnostic reasoning unit. The entire system, however, functions as follows. The knowledge engineer translates the expert's wisdom into a form that can be accepted by the knowledge interface for compilation. The compiled information is then transferred to the knowledge base, similar to a database, where it is now ready for use in the reasoning processes.

The UUT, or unit under test, is a device of some combination of hardware and software that is under current scrutiny. It has a technical description that includes normal operational conditions, as well as abnormal and/or failed situations that may befall it. These various conditions, both normal and abnormal, have indications as portrayed by output responses to input stimuli. These responses are maintained in a database and used as the basis for judging outage conditions as real-time or live inputs are recorded. The fact base and the knowledge base are both inputs to the reasoning process where conditions are examined, diagnostic paths are identified and pursued, and judgments are made about specific components suspected of causing the problem(s) at hand.

7.9.3   Recap of Troubleshooting Issues

Self-Contained Computer Program

The most popular and common approach currently used for troubleshooting is that of the self-contained computer program that incorporates the troubleshooting algorithms and the norms at which the system operation is aiming. As indicated earlier, the current approach is to implement a cumbersome computer program that contains the analytical capabilities, the database required to support the problem domain, and the user interface facility. This method has serious drawbacks, mainly due to the fact that the program is inflexible, difficult to modify, and lacks modularity.

Diagnostic Engine or Reasoning Environment

Another issue concerns that of the diagnostic engine. Up to now, this has been the rule-based inference engine. The rules used by this engine are of necessity only as good as the expert and the person who translates such knowledge into a machine-readable form, usually the knowledge engineer. Additionally, these rule sets may have gaps, dead ends, conflicting elements, etc. Future diagnostic reasoning engines will be designed so as to overcome these inherent problems.

There are a number of trade-offs to be considered in deriving a knowledge acquisition system. At one extreme, we may consider writing a troubleshooting decision tree that contains all possible paths for failure within the UUT. In a slightly more sophisticated approach, a decision tree that guides the actions of an automatic test equipment (ATE) unit is sometimes implemented. Using either method, the burden of diagnostic reasoning is upon the technician, because he or she must determine the paths to be pursued, and results are directly dependent upon what he or she enters into the computer system containing the troubleshooting algorithms. Either way, the effort in coding the decision tree is minimal; however, the cost of generating a decision tree to begin with is considerable, because of its labor-intensive content.

The technician must also consider many issues at each point in the decision tree. For instance, the technician must keep track of which paths have been tried, which modules are currently suspected of being at fault, what tests will rule out different modules as culprits, what short cuts, if any, will minimize the test-suite time consumption, etc. This method, therefore, has high knowledge acquisition costs, and even higher costs for operational usage due to the training required of the technician and the time required to actually run the process to conclusions.

At another extreme, is the approach where the UUT model, i.e., actual detailed schematics of the device, are captured within the computer system, along with the relevant behavioral, i.e., functional characteristics of the device under test. The trick here is to deduce enough of the UUT behavior in order to isolate the offending modules. This approach requires that some human specify the intended function of the UUT to the computer system in order to allow isolation. Another problem with this method of approach is that the diagnostic reasoning system must also know how much of the UUT's behavior has to be tested in order to certify that the UUT has completely acceptable performance. This method has low acquisition costs, but high functional analysis costs to include all reasonable variations pertaining thereto in order for the system to be capable of the certification step.

The features of these new engines will be able to structure their own rules, and collapse them as necessary for dictating circumstances. Such engines are referred to as causal expert systems. These processors will also be capable of picking paths based upon the structure of the system, and picking causes for the information needs. One of the big challenges for these new breeds of diagnostic processors will be the description of the behavioral knowledge of the devices involved. The physical and functional knowledge of operational units is what this issue is about. Whether these data are self-organized or imparted to the system by an outside, human-originated source, their importance to the value of the expert system is critical. A current form compromise of the issues discussed above is presented in the next section.

7.10   Diagnostic Reasoning Optimization

A near-optimum compromise for the diagnostic reasoning system of today, with our present limitations on technology and thought, is provided here. There are three sets of knowledge that must be available for this compromise system, namely: (1) a simplified model of the qualitative, i.e., functional, description of the behavior of all relevant replaceable units within the system; (2) a structural, or connectivity, relationship model of the system; and, (3) a specification of the performance of the UUT. These tools now provide the reasoning processor with the capability to deduce faults from operational abnormalities.

The diagnostic reasoning component can now determine relationships between test results and possible faulty modes of operation. Causal rules can also be applied to deduce chances for any replaceable unit to be at fault given certain observed outputs and their association with internal module functions. This process is reliant upon failure statistics that sometimes are difficult to derive and/or maintain. The key to success for this type of diagnostic reasoning system is to isolate replaceable modules to the extent possible.

The advantage of having failure statistics available to the diagnostic reasoning processor is illustrated here. If, for example, we have a set of modules among which are M1 and M2, and the elimination of one implicates the other as the culprit of a functional failure, then we proceed to uncover the relative health of each module. We begin by trying to execute a test that involves one module without involving the other. Suppose we find such a test for M1 and it fails. The implication now is that M1 is at fault, and M2, forgetting that both could also be faulty, is exonerated. Many times, however, testing is not a black or white circumstance. The judgment of what is a passed test and what is a failed test is many times open to interpretation. This is the point at which joint probabilities and single-point probabilities of the modules in question are of potential importance.

7.11   Calculating Fault Probabilities

Fault probabilities are a big part of diagnostic reasoning systems, such as MYCIN and PROSPECTOR. These reasoning systems incorporate probabilities to make judgments about the evidence collected, and conclusions reached. In the medical domain, it is much more difficult to assign probabilities, possibly because joint statistics are not available for use. In the case of engineering environments where the system is constructed from discrete components, statistics, and in particular joint statistics, are more easily come by. Precise probabilistic methods can be derived and used. A failed test in a medical environment may not yield any information, because of the variability of the conditions under which the test is run, but in a human-engineered system, a failed test can be more significant. A failed test may not isolate the failed component, but it may restrict the range of ambiguity to a small number of units.

The effective calculations of fault probabilities are based upon the notion of fault hypotheses that can be formulated by using mathematical dependency rules. Thus, probabilities of joint events where the probability of any one event is independent of the other, are calculated by multiplying these event probabilities together. For example, the probability of drawing any ace in a deck of cards, and then drawing any heart, given that the ace drawn was not the ace of hearts (without replacement), is:

$$\text{Probability (ace, heart|ace <> heart)} = \left(\frac{4}{52}\right)\left(\frac{13}{51}\right)$$

This probability is approximately $\frac{1}{50}$, which is roughly the same as drawing one specific card, such as the 3 of spades. So, a fault hypothesis is a boolean combination of probabilities, each of which is specific to its own component. This combination might be expressed as $P_{1_f} * P_{2_f} * P_{3_g}$, where we are interested in the hypothesis that module 1 and module 2 are failed, while module 3 is operative. The diagnostic reasoning system should have a built-in facility to calculate such hypotheses based upon the intermediate test results that are being gathered.

In the scenario discussed above the components are assumed to be independent of each other, such that the failure of one does not impact the operability of another. Thus, here we are ignoring the case of one unit's failure affecting the failure in another unit or masking its operability so that it appears to have failed. Therefore, double faults are still independent events. Once the procedure for calculating failures has been determined, we have to determine the strategy and/or the next troubleshooting test.

Initially, troubleshooting is a random process, becoming more and more specific as tests and results begin to build up a body of evidence pointing to one or more conclusions, i.e., failures. Any of the previously discussed methods, e.g., fault trees, semantic nets, or causal approaches, can be applied to further this game of fault isolation. An actual test consists nominally of three tasks, namely setup, connection, and measurement. The initial test is heuristic, i.e., the test is conducted that maximizes the ratio of information gain to test cost. Information gain is a weighted average of evaluation functions for various several possible test results. The weights for the weighted functions are the test result probabilities. The computation of the evaluation function is provided as follows:

$$\text{Evaluation Function (EF)} = \sum_{i=1}^{2n} p_i \log p_i$$

where: p_i = current fault hypothesis probability
           n = number of UUT modules

Unfortunately, the number of calculations required to evaluate all possibilities of n components can be quite large. For instance, 10 components alone require 1024 separate calculations to assess all possibilities. This is why some form of heuristic must be developed to limit the range of calculations.

Summary:

Expert maintenance systems (EMSs) can justify themselves on the basis of cost savings, or cost control. The key to effective EMS development is the extent to which sensor control is possible within the target environment. The more pervasive and/or intelligent the sensors are, the less inference is needed in the EMS to isolate and identify the source of the operational alarm(s). One of the major issues with EMSs is their susceptibility to ambiguous findings. Such problems can be mitigated through the employment of smart sensor systems that aid in isolating offending components and, therefore, help reduce ambiguity issues. Network management environments are made to order for the utilization of sensor placements, the collection of sensor data, and the analysis of this data for the purposes of diagnosing the maintenance problem at hand.

<u>References</u>

(1)   Cortner, J.M., *Digital Test Engineering*, John Wiley & Sons, New York, 1987.

(2)   Fischer, M. A., and O. Firschein, *Intelligence: The Eye, the Brain, and the Computer*, Addison Wesley, Menlo Park, CA, 1987.

(3)   Hopple, G.W., *The State of the Art in Decision Support Systems*, QED Information Sciences, Inc., Wellesley, MA, 1988.

(4)   Kiggans,R., et al., *Expert Systems*, Education Foundation of the Data Processing Management Association, 1986.

(5)   Loomis, M. E. S., *Data Management and File Structures*, Prentice-Hall, Englewood Cliffs, NJ, 1989.

(6)   Muller, N., and W. Victoria, "The Promise of Expert Systems," *Procomm Enterprises Magazine*, Dec/Jan 1990.

(7)   Richardson, J. Jeffrey (ed.), *Artificial Intelligence in Maintenance*, Noyes Publications, Park Ridge, NJ, 1985.

(8)   Scott, P., *The Robotics Revolution*, Basil Blackwell, New York, 1984.

(9)   Strauss, P., "Network Management in the '90s: The Issue Is Intelligence," *Business Communications Review*, June 1990.

(10)   Wallach, R. M., "The E.I.S. State," *Computer Systems News*, March 3, 1990, p 49.

Chapter

8

Expert Systems: Performance Systems

Chapter Highlights:

Expert performance systems hold the promise of providing continuous oversight of the operation of the telecommunications equipment or network. Such requirements as the network's operational efficiency and traffic management are sufficient to justify the development of expert-based systems for performance monitoring and decision making. The typical network consists of functional control, database systems for data storage and retrieval, and performance monitoring features. This latter activity is the basis for measuring the effects of planned actions and provides the criterion for corrective actions. This chapter explores the various attributes of the expert performance system, its design, and its place within the existing architecture of the typical network management system. Operational performance monitoring is also thought of as a tool for maintenance requirements.

8.1 Overview of Current Operational Systems

Maintenance systems have received considerable attention over the past few years, primarily due to the fact that they had tremendous potential for justification through cost savings, and because they were more easily implemented in the constrained environments of the system domain. Soon, however, attention began to be directed toward the operational aspects of the systems. The issues of monitoring system operations and measurement of operational performance emerged as key parameters of the overall system effectiveness evaluations. These system development efforts emerged in several different areas that are summarized below. As systems become more complex, performance measurement and monitoring take on a higher degree of importance as the watchdogs for the network management organization.

(1)   Databases

The database is the single thread between all the operational and control features of the operational system. It provides not only the linkages between operational and geographic nodes but also provides the data within which these entities operate. The performance management function relies upon the databases used in the system, and therefore expert system approaches to the control of these entities are a key activity in this area.

The database system has two major attributes, namely to identify what parameters are important for monitoring and to track the values of these parameters as the system operates on a continuous basis. The design of the database is important for the extraction of relevant information as needs require and to associate this to data with other parameters so as to make decisions about system operation with higher levels of confidence. This design usually involves the use of architectures that support distributed, relational features. With network databases being implemented within the confines of telecommunications networks, the distributed features of such database systems become very important. The distributed database can be utilized to support sophisticated analyses about the nature of performance or fault deviations, such as: (1) trends in performance; (2) the changing status of associated equipment or components, to assess the cause and effect of such changes; (3) usage sensitivities; and (4) physical or functional changes in the equipment that may affect performance indicators.

(2) Functional Control

Once expert maintenance systems have diagnosed or isolated the fault(s) with which they are currently involved, there is an additional task that must be performed, namely the dispatch and management of the repair or alteration of the dysfunctional situation. This control feature is part of the operational system performance management function.

(3) Performance Monitoring

One of the primary areas of importance in system management is that of performance monitoring. Another area closely associated with performance monitoring is that of traffic control. Its function is to monitor the existing calling patterns of the network and to recommend alternate routing for such loading issues.

This chapter identifies these areas of performance management and explores the expert systems development activities that have marked their transition into the AI environment.

8.2 Intelligent Database Systems

8.2.1   Problems

Intelligent database systems address several shortcomings that are inherent in the existence and use of multiple database systems. These shortcomings include:

(1) lack of a natural language user interface
(2) common user menu options
(3) lack of interactive scenarios and user profiles
(4) lack of a standard control and format for the presentation of
      data
(5) absence of an understanding of the boundaries of the database
      knowledge
(6) accesses to multiple databases
(7) data combinations and fusion
(8) query programs
(9) communications with other databases

Functionally, these problems are attacked from two directions, the front end and the back end. The front end provides a natural language user interface, standard menu selectivity, controlled usage scenarios, the implementation of user profiles, and the control and format of data presentations. The back end provides expert system database functions, such as domain understanding, multiple database selection, data fusion from different databases, query program generation, and distributed database communications.

These problems are so constraining that an effective distributed communications system, let alone a database system, required some innovative actions and redesign efforts. Shared and multiple database systems technology took a turn when networking became the preferred method of using computer-based systems. A whole host of new capabilities had to be developed in order for this new use to be realized in widespread and common form. These new technologies included physical connectivity, communications protocols, compatible operating systems, resource sharing, and many other facilities. With the rapid decline of computer power and the rapid increase in the magnitude of that power, it became clear that time sharing per se, where one mainframe would serve many terminals, was no longer an acceptable alternative for computing organizations requiring more and more local control over the applications and the attendant sharing of database resources.

The next several steps in this evolutionary process yielded a network environment where many computers were tied together within an environment that also supported storage, printing, and outside interfaces, e.g., modems, and other services. The database requirements of these environments soon reached severe limitations. The basic methods by which multiple databases communicated were by transmitting messages between each other, or by using rendezvous, where messages can be left by one database and retrieved by another. These approaches did not work adequately because the access and updating procedures were not up to the challenges. The distribution of databases arose because of geographic, functional, and organizational needs, to name a few.

Traditional database systems also support only fixed-format query languages that are targeted solely for one database system at a time within the database management system under consideration. Such systems were not intended to interact with other database systems or even databases within the same system.

8.2.2   Functional Requirements
for Intelligent, Distributed Database Systems

A detailed functional requirements definition and analysis of the typical distributed network database operation is difficult to perform, because the requirements are undefined for a specific case, and tend to be variable depending upon the circumstances of the situation under investigation. However, a representative set of needs can be assumed for the purposes of the discussion here, and their implications explored. The data needs for the current-day user of distributed databases include concurrent access to two or more databases, of either similar or dissimilar varieties, simultaneous access of two or more people to the same data set, and data set updates during multiple access periods. Each of these will be related to intelligent systems (IS) approaches as described below.

The concurrent access of the user to two or more databases at the same time presents several IS problems. First, an IS approach would entail the use of a consistent user interface that would incorporate a natural language processor and standardized menus, as well as user and scenario profiles. Additionally, the IS would provide a data fusion capability that utilized data deduplication, data domain understanding, and some inference features based upon the combined data sets available.

The issue of simultaneous access to data by two or more people, assuming they are all authorized to such accesses, brings up the problem of data updates, and how they are to be handled. Such a problem can be handled by the IS through the use of interaction scenarios, user profiles, and data domain understanding. The IS could be configured to monitor the condition of the entities accessed, e.g., files, records, or attributes within some record, such that an additional access can be recognized, tracked, and any changes noted for other simultaneous users. Figure 8-1 shows the generalized functional needs associated with the intelligent database requirements. The functional requirements could be further subdivided into three categories, namely user interface modules, special-purpose utilities, such as formats, and the expert systems modules. The situation as it pertains to each function required or desired within these categories is explained as follows. Figure 8-2 shows the general architecture for a distributed database system that would rely upon the functional requirements discussed here.

(1)   Natural Language Processor

This user interface has already been discussed and can be dismissed here except to say a word about its fit within the functional requirements environment. Natural language interfacing is an important topic and one that can make the difference between a useable tool and one that is a key to successful operation. This makes natural language processing a function of importance on its own as opposed to being a task that is dependent upon another activity for its full level of importance. The objective here is to make the user's interaction with the system as simple and natural as possible so that operational needs can be satisfied with minimum efforts.

Figure 8-1
Functional Requirements for
an Intelligent Database Management System

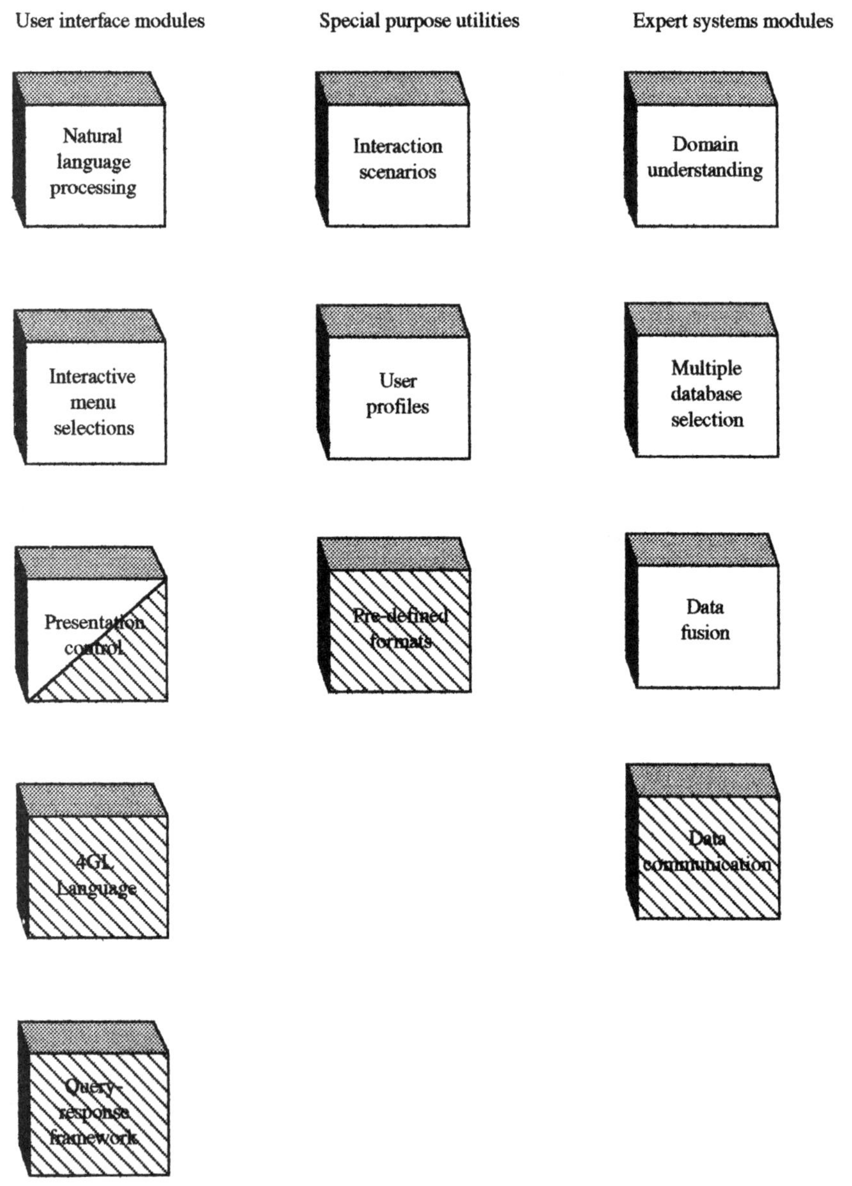

(2)   Interactive Menu Selections
The menus determine the ease with which the user can maneuver within the system. One of the prime user interface functions is to provide not only natural communication with the system, but also simple, easy-to-use menus.

Menus have several forms, among which are those that are part of some data presentation and those which are not. The latter variety include those menus so common to us all where the menu occupies the entire screen and is a filter through which the user must proceed before passing this particular gate. In other words, a selection must be chosen in this type of menu prior to the menu giving way to another presentation, either another menu or a formatted presentation of the requested data. The former type of menu includes those menus that are embedded within the current data presentation. This can be provided in a couple of popular ways, one being the now familiar pull-down menus that are part of many operating systems, such as Microsoft Windows, or the Macintosh environment.
The other method of menu selection incorporates those types that are embedded in the information bars of the presentation. In this approach, the menu selections are constantly presented on the screen at either the top or bottom of the presentation within menu bars. Each menu bar can be selected to reveal a pull-down menu of items under the heading indicated by the menu name, such as *file*.

(3)   Presentation Control
This function has both a standard and expert basis to it. Its standard operation is its requirement to provide a consistent presentation format and protocol for incoming data so it can be presented in a similar form. Its expert aspect embodies its ability to recognize differing input data streams and to reformat these data streams on-the-fly, i.e., as they come in from the external environment, so that they can be integrated into the desired format presentation.

(4)   Fourth Generation Language
The fourth generation language (4GL) environment, as it is called, is a nonintelligent feature of the database requirements list. Some sort of interface language is a standard feature of the normal database system. This function is one of providing a facility for the user to communicate effectively with the database so as to extract the desired information in the desired presentation format.

(5)   Query-Response Interface

Some type of query-response capability is essential to carrying out the basic requirements of the standard database system. It facilitates the collection, retrieval, and commanding of the database with which it interfaces. Query-response facilitates the on-line interactions between the user and the computer system.

(6)   Interaction Scenarios

An intelligent systems environment requires that some form of feature be incorporated that accommodates different situations that the user may specify on an a priori basis. Such invocations may allow the system to set up data integration paths from several sources, self-query capabilities, and/or special referencing permissions where the system may, on its own, request additional data from sources so as to complete the picture desired by the user. These interactions are simulations that provide the user with what-if data needed to answer questions and to make decisions.

(7)   User Profiles

The user profile is a set of information from which the intelligent system may draw in order to make quasi-independent decisions about the processes described above. Such issues as user intents and limitations imposed by the user may be incorporated into this intelligent database function.

(8)   Predefined Formats

In many situations predefined formats may serve the needs of the situation adequately. However, in a growing number of scenarios, the formats of the displays or outputs may not be capable of being identified beforehand. In these situations, a facility to format data on the fly would be of value to the user. In these circumstances, the user profiles and the interaction scenarios would be used to derive the user's preferences for data presentations.

(9)   Domain Understanding

For the expert systems environment, it is critical to define and work within the nature and extent of the domain of operation. This the meaning behind the term *domain understanding*. First, the extent of the domain is that boundary around which the expert database system operates.

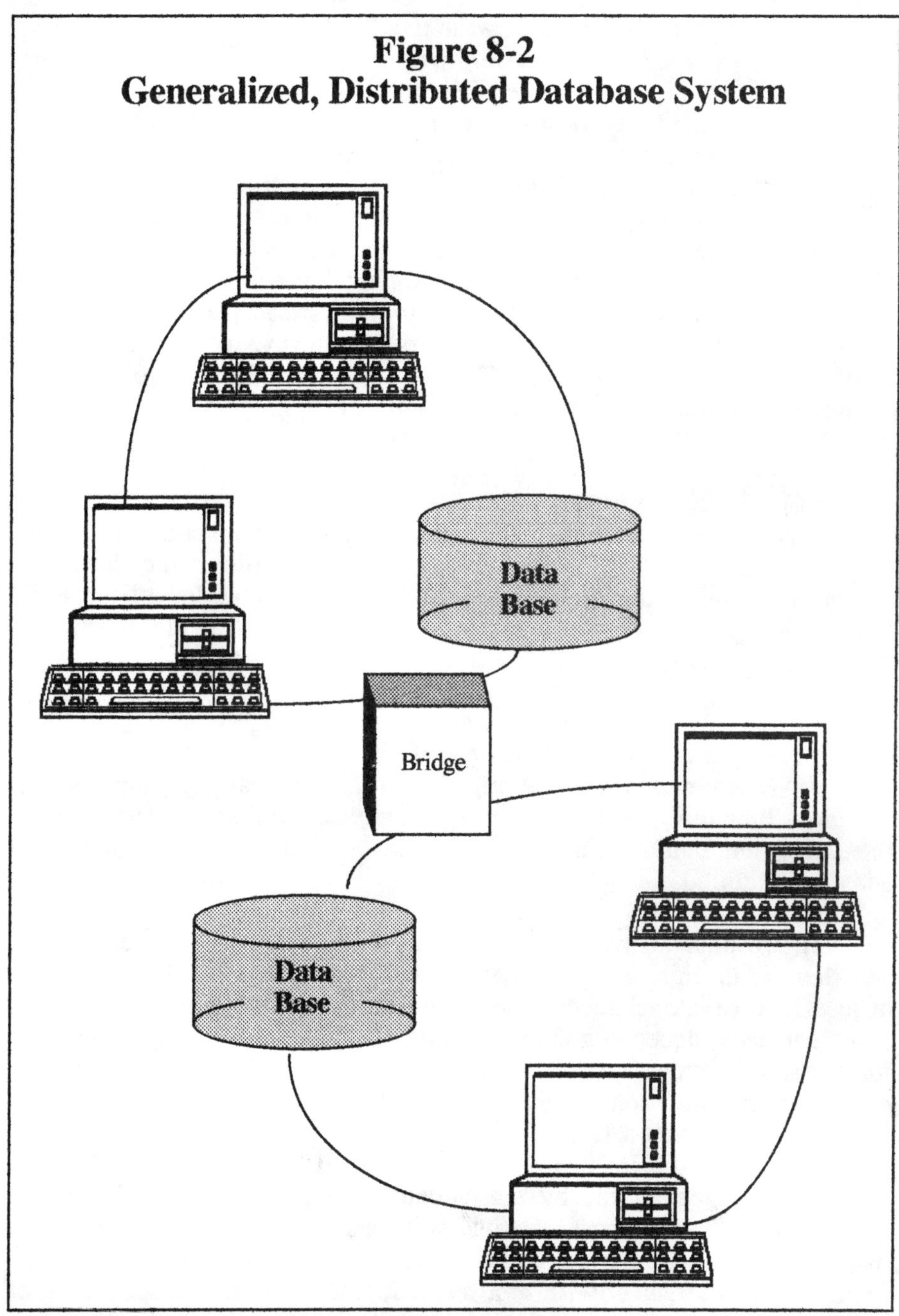

Figure 8-2
Generalized, Distributed Database System

Second, the nature of the domain is that definition of the domain in which the data are interpreted and used by the intelligent system.

(10) Multiple Database Selections

A key function in the expert systems environment is that facility that allows and performs the queries of multiple databases. This facility is performed by the multiple database selection function. There are two types of such queries, one being multiple requests to one database from several databases within the network, and the other being multiple requests from one database to many databases. In the first case, servicing requests are processed on the basis of time of arrival, priority, request source, etc. In the second case, the requesting database waits for responses to queries to be filed before completing the processing transaction.

(11) Data Fusion

A key function of the data manipulation functions is that of data fusion. The expert systems intelligence that identifies and collects data for the user's needs performs a data fusion function that deduplicates and fills in data gaps as appropriate for the specific needs of the question at hand.

(12) Data Communication

A nonintelligent systems issue is that of data communications. For those functions that need to identify and access different data sources, the intelligent modules in those functions will perform the connection activities required.

If two or more workstations are trying to access the same database and there is no dual access feature in the hardware, a blocking situation arises. If there is dual access, the same data can be accessed and retrieved simultaneously, depending upon software constraints restricting access that may exist at the time.

For the situation where two or more workstations are trying to update the same data set, the circumstances become more complicated and give rise to the problem that has created most of the stir about distributed database management and the subsequent expressions of need concerning intelligent systems support. Notwithstanding the fact that most databases have security restriction capabilities that allow limited accesses to files, records, or attributes within a record, the generalized problem of simultaneous updates and propagation of those updates to current users of that data becomes an important topic of concern to the user. The fatal flaw, however, for simultaneous accesses and subsequent updates of the item in question is the situation that occurs when one user accesses and updates an item after a second user accesses the same item, after which again, the second user updates the item in question that was previously updated by the first user. The consequences are that the first user's update is lost, and subsequent accesses and updates by whomever will propagate the omission of the missing updates.

The possible scenarios involved in the dual access issue can be characterized as follows. First, a current active access could restrict another access request until the first access is completed by some sort of database flag or active signaling protocol back to the new requestor. Second, a data access could be allowed for multiple requests with the constraint that any update is handled in chronological order and not propagated back to other current users, or is propagated, as the design dictates. Third, concurrent data accesses could be allowed with updates held in temporary records until the master database is updated as a whole at specified points each day or other time period. Each of these implementations may be driven by the application required, or the specific database software design implemented.

8.2.3   Intelligent Database Architecture and Implementation

The basic idea behind the intelligent database approaches is that these facilities interface with the normal databases to yield the data needed, in the right format, and utilizing the inference techniques of AI to enhance the data with more meaningful implications. The actual implementation of the intelligent network database system can be approached using the notion that any circumstance can be captured and stored for reference. This method is referred to as the frame-based approach. The data structure architecture for this frame-based approach is illustrated in Figure 8-3.

Theoretically, any of a system's features can be reduced to some sort of database description that reflects the physical, functional, or operational peculiarities of the system in question. The key to the intelligent systems approaches implemented will be the efficacy or the working integration of conventional databases and intelligent shells that attempt to interact with these data structures. Figure 8-4 shows the basic intelligent database components of the typical network.

The frames created from operational data inputs can be used in various configurations, or flavors, such as physical, functional, etc. For instance, a physical frame might represent a normal configuration for a particular type of operation. This frame could be used as an overlay to advise the user about necessary connections needed to be made in order to satisfy the requirements of such an operation. Alternatively, an operational frame might contain the parameters that need to be monitored in order to satisfactorily track the progress of a particular configuration needed for the operation specified. These components are applied to the network in such a way as to maximize functionality and to minimize cost impacts.

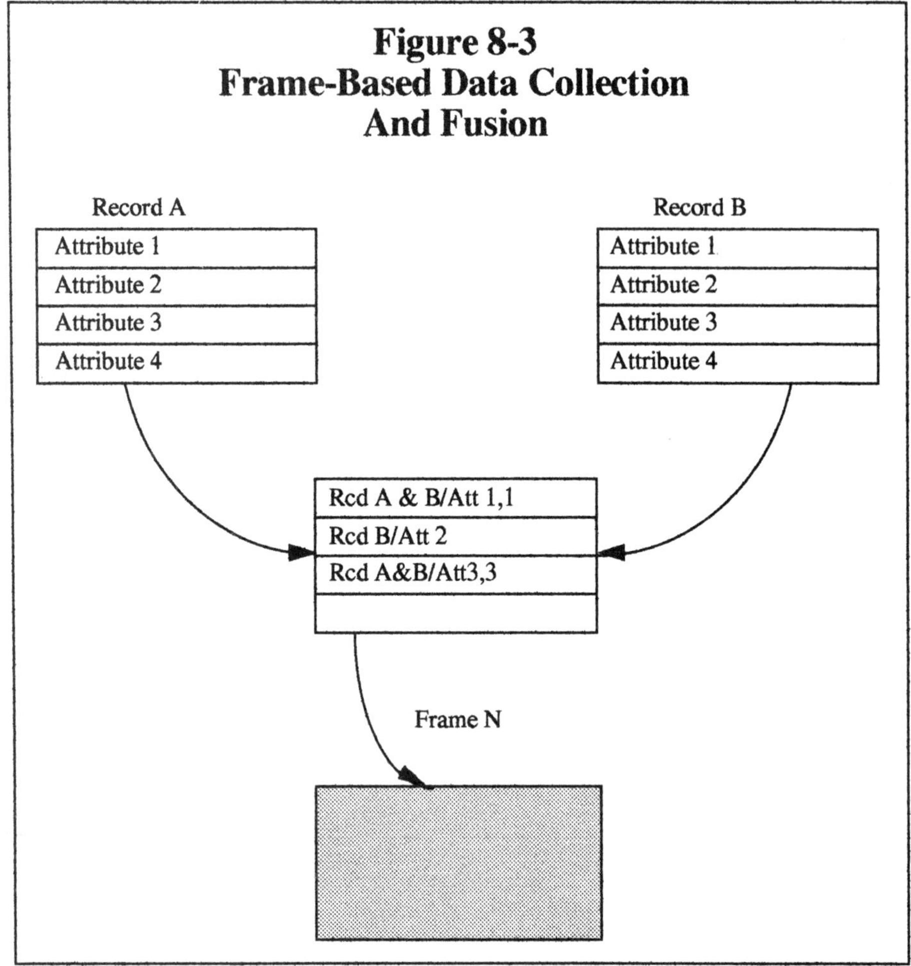

The shell interface for the intelligent database system is further reflected in Figure 8-5. The shell operates as the interface between the external environment, as depicted by the database system, and the user. The communications facilities provide that link between the internal and external forces in action. The expert systems shell can control the database and communications systems so as to impart intelligence to them by having these entities respond to queries at appropriate times for data collection with which the expert system interacts. The intelligent database feature of the system can assemble data frames and retain them for use by the user based upon the user's profile of information needs. The intelligent data server can also query databases from various sources and assemble these diffuse sources into a fusion of data needed by the user as well.

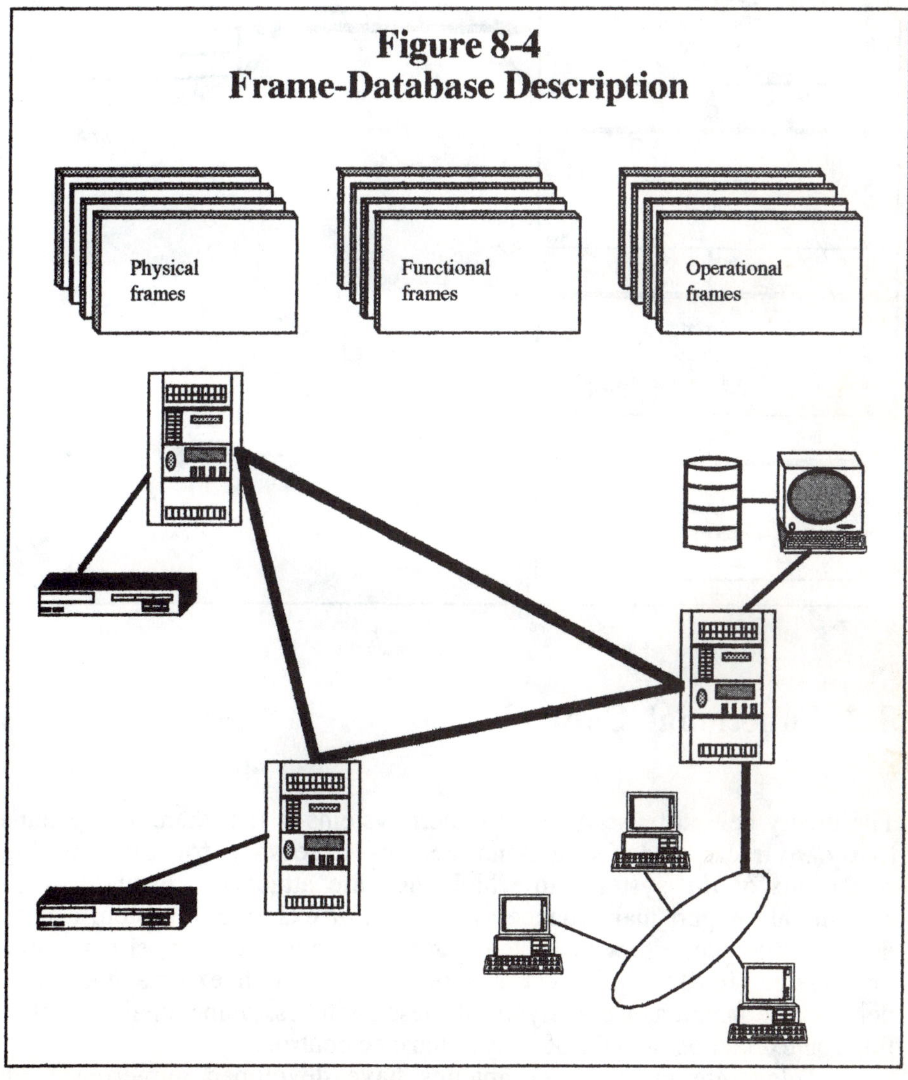

**Figure 8-4
Frame-Database Description**

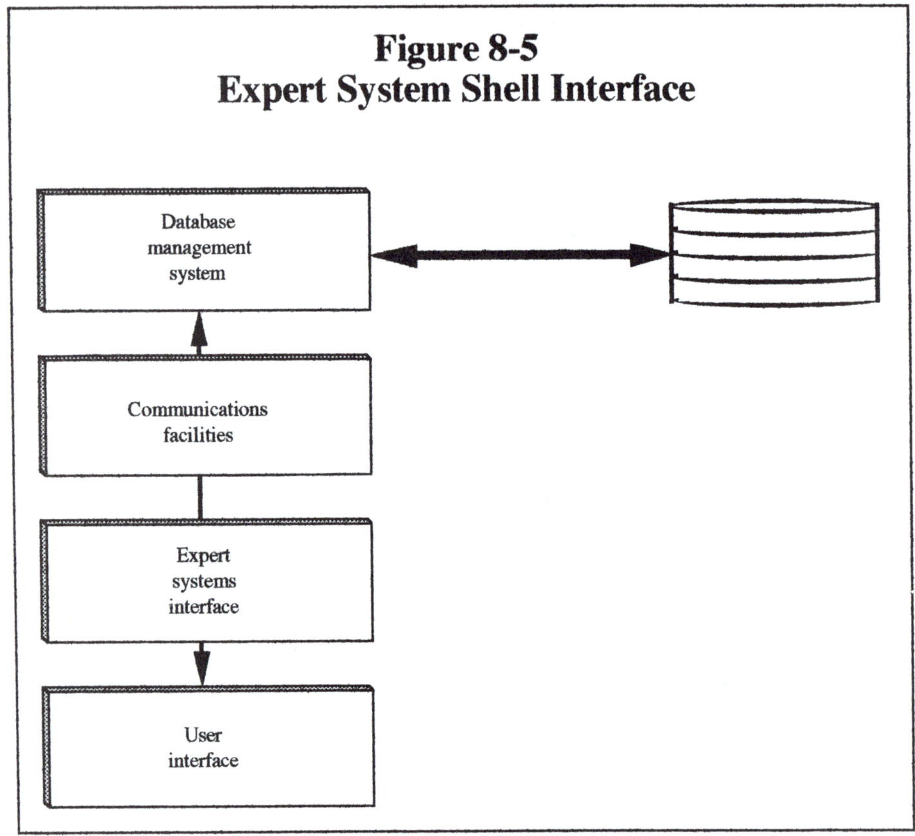

8.3 Functional Control

The theory behind these types of expert systems is that there are systems that can track, and recommend actions necessary for the ongoing operations of the systems to which they are attached. In short, such systems are expert maintenance systems. For example, a system can be developed to support switched telecommunications networks, either public or private. It can replicate the expertise of switch experts needed to perform the maintenance analyses of these switches. Functional control in this context can be thought of as maintenance control.

Telecommunications companies have developed numerous such systems over the past few years. These systems have been directed mostly at automating the now manual maintenance activities required to keep these complex technical operations going. For public, or even private, switching systems, large numbers of status messages are generated every hour. For some systems, expert capabilities have been implemented to analyze each error, fault or abnormal indication and recommend the actions that need to be taken, the priority of such actions, and the justification for these recommendations. It is not uncommon for the diagnosis to target faults down to the circuit card, switching relay, or wiring pin level, all of which can be classified as line replaceable units (LRUs). Figure 8-6 illustrates this facility.

Each switch has a status reporting segment whose output messages can be relayed to a central facility for analysis. The volume of such messages dictates that either a preprocessing be performed at the point of origination, or that this task be performed at the receiving end to cull out duplications, crossover messages (those that point to the same point of initiation, but come from two or more different sources), repeat messages, etc. Once this has been accomplished, the received data messages are formatted so as to facilitate the retrieval of these data in formats suitable for the intelligent back-end programs to manipulate. As discussed earlier, such formats might be data frames to capture the physical, functional, or operational views of the system under scrutiny.

The expert system, at this point, fuses the data from diverse sources to generate a picture of the situation at hand. Data fusion performs the task of increasing the confidence level of the reported errors by supporting the assertion with corroborating reports. It also adds information that would otherwise be absent. Since the overall process is fuzzy, or probabilistic at best, several possible outcomes are likely to be presented and associated probabilities assigned to the likelihood of each occurrence. The data fusion approach allows the expert system to react with more knowledge that can be used to examine the rules, frames, etc., for the purpose of reaching the best conclusions, and to do this more efficiently than might otherwise be the case.

The user interface consists of a natural language facility that can interpret requests, and a variable output capability associated with user and scenario profiles similar to those discussed earlier. The most common output might be maintenance recommendations of certain configurations, coupled with the priorities and justifications for each. These additional features give the interested user the ability to examine the logic and reasoning used to reach such conclusions, so that differing objectives or conclusions can be resolved by user override, for example.

A specific set of actions that might be required of a functional maintenance system are indicated below in their order of occurrence. The steps presented below actually come from a system developed by GTE, called **COMPASS** (Central Office Maintenance Printout Analysis and

Suggestion System), that was developed to analyze just such switching system reports and faults.

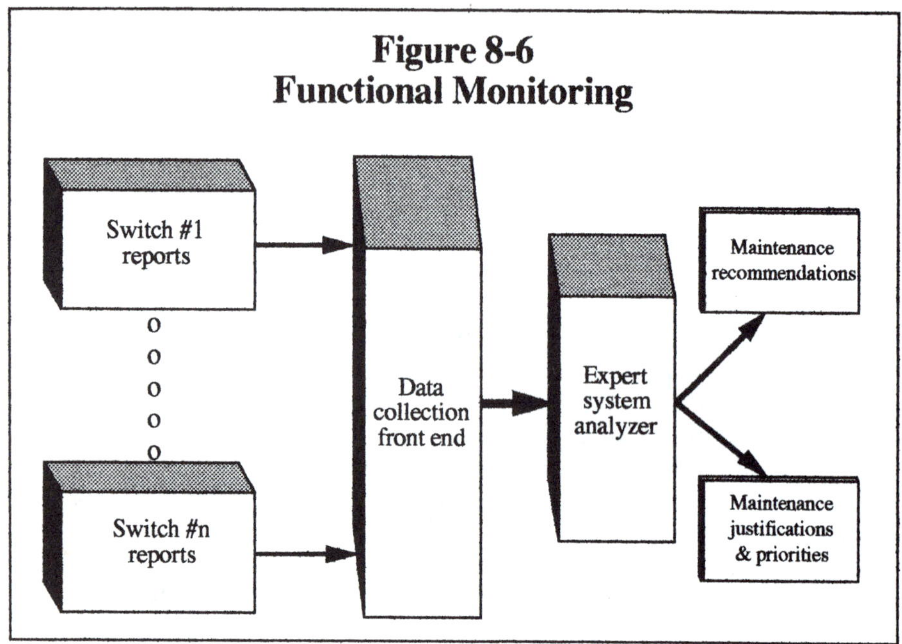

(1)   Maintenance messages are grouped so that those associated with a particular switch can be examined for duplicates, etc.

(2)   Specific message groups are analyzed so as to isolate specific faults or abnormal operations.

(3)   The system estimates the likelihood of the occurrence of each possible fault reported.

(4)   The system determines the actions likely needed to resolve each fault reported.

(5)   The system prioritizes the actions required by considering the likelihood of each fault, its ease of action, performance impacts, and possible risks of each action taken.

(6)   There is a merging of different actions on the action list thus derived so as to combine maintenance actions that can be performed in concert with other maintenance actions of similar nature or those that have similar points of application.

(7)   The system provides an output that is most familiar to the personnel viewing or reviewing the recommendations reached by the expert system. A maintenance operator would probably want the output presented in a form that resembles a to-do list, while a supervisor might want a pictorial representation of the actions recommended so as to visualize the overall picture of the work to be performed.

Although not part of the current COMPASS system, but of possible importance, there are some additional ways of assessing the impacts of maintenance actions upon the system operations, and the system's normal maintenance scheduling and backup capabilities during the periods of maintenance actions. For instance, if the system maintenance actions adversely affect standard operations, then it would be helpful to have a facility to schedule the maintenance work so as to dovetail with regular maintenance, and/or to create a system downtime situation during periods of anticipated slack demand. Of additional importance is the need to assess the backup capabilities of the system during these periods so as assess emergency capabilities available.

8.4 Performance Monitoring

Under the category of performance, there are several subcategories that could be discussed, such as operational efficiency and traffic management. Operational efficiency is that level of activity that monitors and measures the overall or specified areas of interest within the system. Traffic management tracks and reports on congestion and possibly makes recommendations about congestion within the system.

8.4.1   Operational Efficiency

The successful operation and maintenance of a system requires that measurable objectives be defined for that operation. It is, therefore, somewhat subjective as to what these objectives are and to where their importance is placed in the overall accomplishment of the system's objectives; i.e., priorities must be established. But, once stated, these objectives should be quantified and the system must report upon these measures so that the objectives can be evaluated for conformity with these data. The objectives and their priorities, once established, can now give way to a definition of the parameters that constitute these objectives. Such parameters might include the following items:

(1) equipment operational availability times

(2) equipment operational maintenance measures

(3) operator interface measures, such as connect times, usage times, and operator dead times

(4) system data interface measures, such as data volumes coming into or exiting the system, frequencies of data exchanges, time critical aspects of these data exchanges, etc.

Operational efficiency measures can also be utilized to define functional outages that would be created by such problems as blockage and overload. Functional outages can be as destructive as physical outages, thus, the measures defined for efficiency may also be categorized as critical parameters for review whenever their values exceed certain desired ranges or amounts. Other uses for these parameters involve the subject of traffic management. Here, the values obtained may be used for problem avoidance before these values exceed operational tolerance levels.

8.4.2   Traffic Management

Traffic management addresses system measures that are tending to become unmanageable, but prior to their actually reaching that point. The timing for traffic management reports is typically every 5 minutes, and alarm status reports as they occur, with summaries provided every 30 seconds or so. The subject and study of traffic management issues has culminated in reducing the management task associated with traffic management to a relatively few equations that are aimed at sizing a network's call-carrying capacity, based mainly on measures of the average, peak, and busy-hour traffic.

For many situations, this type of measure does not work, because there is no such thing as an average peak. Christmas, disasters, and other special event communications patterns negate these measures due to their lack of recognition of these extraordinary circumstances. Also, cost-effective habits dictate that the systems not be designed for the absolute worst imaginable situation, but rather an expected average peak scenario.

As a result, traffic management is an important issue that must probably remain partly manual and partly automated.

Traffic Failure Modes

There are four typical categories of problems facing traffic managers. In all such cases, the demand, for whatever reason, exceeds the capacity of the system to meet that demand. The experience surrounding these capacity problems was mainly gained over the years from the telephone system. The definitions and categorizations of overload in telecommunications are, therefore, associated with the telephone system. Each of the situations presented below is differentiated by the source, extent, or alternatives available to the problem solver.

(1) Routine Capacity Exhausted

Telecommunications facilities are designed for expected traffic increases over some specific period of time. Quite often, these design growths are exceeded by unexpected circumstances. As a result, certain routes become chronically overworked and capacity is routinely surpassed, resulting in blocking, and/or slower-than-average service delivery to the customer. There is a lag between the recognition of such a problem and the resolution of that problem with the increase in capacity of that particular route.

(2) Focused Overload

This type of overload occurs when there is sufficient capacity in the network for the load, but localized demand is well in excess of local capacity. This type of problem happens when there are such events as disasters, power failures, or mass call-in events, such as telethons.

(3) General Overload

A general overload occurs when the network capacity is exceeded across the entire network area, and not localized to just one region. The major difference between this type of problem and the focused overload problem is that the overload is so widespread in this case that the alternatives for control and resolution are reduced significantly. This type of overload is characterized by the Mother's Day or Christmas calling demands, but the bigger fear is that it can also be caused by a local problem that is handled poorly. There was an AT&T software problem that occurred in 1991 that propagated throughout its network and caused overload havoc across its entire system, or at least the software malfunction created the perception of a network overload.

(4)   Facility   Failure

In this type of situation, the demand exceeds capacity because capacity has been reduced through some form of failure mechanism. Usually, the failure is a trunk group or an entire cable, but the problem is only temporary. In more drastic circumstances, the outage is caused by some form of natural disaster, such as weather or fire. The Hindsdale fire of 1989 is a case in point, where the entire central office facility burned, creating an immediate loss in capacity. In this particular case, AT&T came in to help resolve the problem and restore the system to its operational capability.

Traffic Analysis Measures

The analysis of traffic must always begin with the work done by A. K. Erlang, a Danish mathematician and pioneer in the field of queueing analysis. Early in the twentieth century, Erlang worked out the equations for several important congestion problems, and thereby earned a place within the discipline of queueing theory. His name has been given to the unit of traffic intensity, the erlang. In general, the traffic intensity of a system, or its number of erlangs, indicates the least number of servers, or trunks, needed to support the system with a given traffic stream at no loss of customers or sessions.

The basic model is one that includes a customer, a queue, and a server. In actual situations the customer is the origination of a communications session, usually initiated by a human or a computer system. The queue is the capacity of the communications system to handle the call initiation load. The queue is actually defined as those arrivals into the system that are being served and waiting to be served. Since the central office has no waiting line capability (i.e., session requests cannot be stacked until a previous session has been released) then the queue is of limited size, the maximum of which is the number of trunks available to the destination. The server, or servers, is the trunk capacity of the system to carry the intended communications, as depicted in Figure 8-7.

Here, one central office handles the customer arrivals based upon its capacity to do so. There are always chances that blocking can occur, which generates a signal to the arriving requests that there is insufficient room in the queue, and that they will have to try again later. The trunk group is the server, and the receiving central office is the point at which the server releases the circuit to either another trunk group for further relay, or to lines that terminate at the customer's premises.

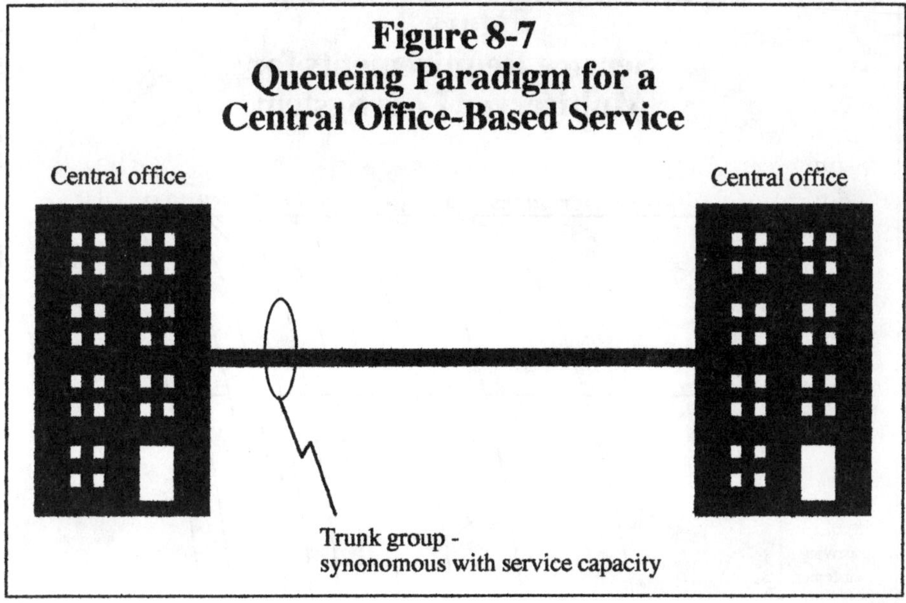

**Figure 8-7
Queueing Paradigm for a
Central Office-Based Service**

We can start the analysis by reviewing the basic single-server queueing equation, which is

$$\mu = \lambda\rho,$$

where:

μ = the system utilization, or intensity, that is equivalent to the minimum number of servers required to provide a no-loss service to incoming customers, i.e., the numbers of erlangs required.

λ = the arrival rate of the customers or service needs, as a function of a distribution that acknowledges some variability in this rate.

ρ = the service rate for each service need, usually a function of some distribution, such as an exponential, random, etc.

This equation is also the starting point for this analysis. The need for multiple, parallel servicing is reflected in the calculated value of μ, which in such cases is greater than 1. We see numerous examples in everyday life of parallel servicing requirements in such places as banks, supermarkets and beauty salons. The value of the parameter μ can be best

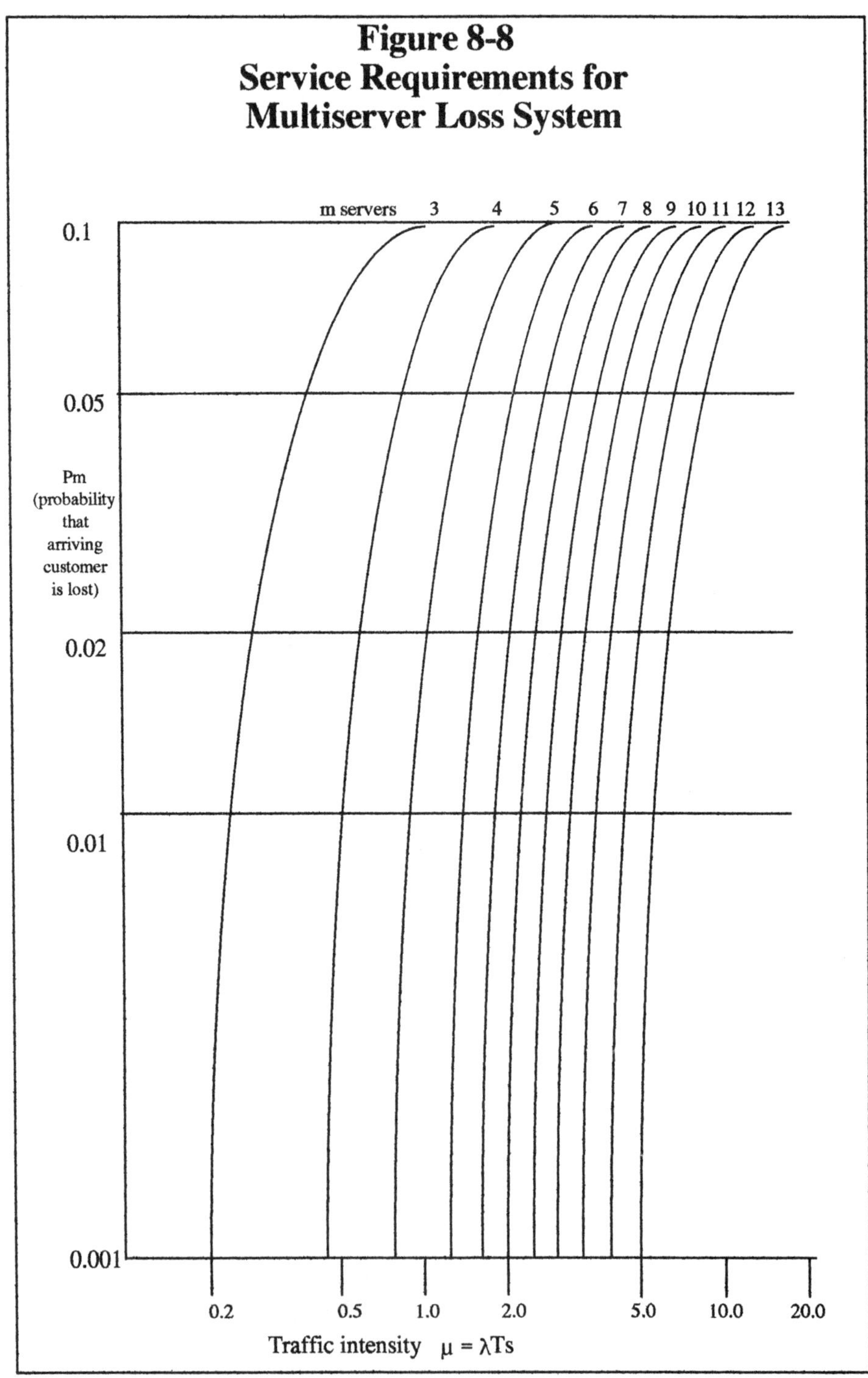
Figure 8-8
Service Requirements for
Multiserver Loss System
m servers   3   4   5   6 7 8 9 10 11 12 13
0.1
0.05
Pm
(probability
that
arriving
customer
is lost)
0.02
0.01
0.001
0.2   0.5   1.0   2.0   5.0   10.0   20.0
Traffic intensity   μ = λTs

derived by examining the requirements for individual pieces of the system's overall need to carry traffic.

If, for instance, there are 10 circuits, each being defined as an intermittent communications session between the two central offices, as in the example above, then the cumulative erlang traffic value is the sum of

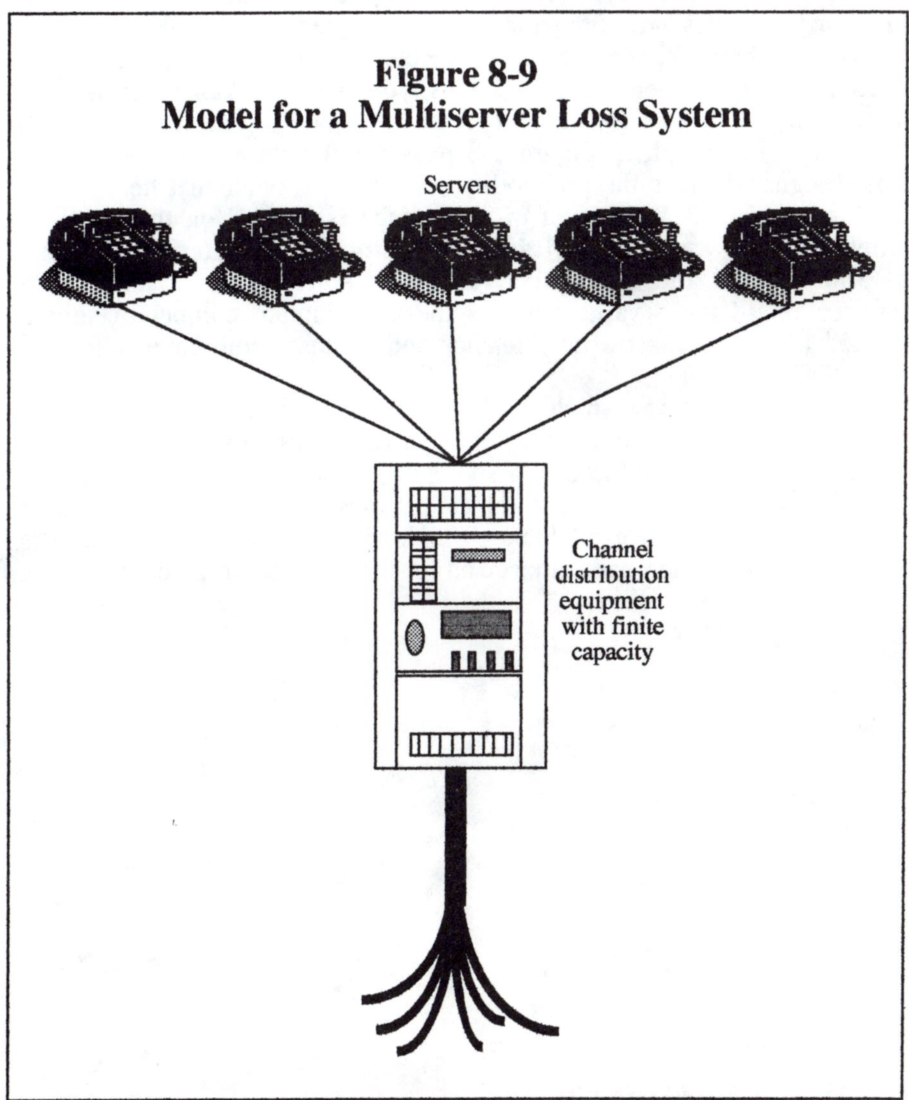

**Figure 8-9
Model for a Multiserver Loss System**

the intensities for each circuit. Thus, if the set of intensities for 10 circuits, using the equation $\mu = \lambda\rho$, is .065 + .098 + .130 + .163 + .228 + .260 + .292 + .325 + .357 + .488, then the value in erlangs for this trunk group is 2.406. This means that three trunks will be required to satisfy the circuit loading of this intercentral office path. The derived value is rounded up to at least accommodate for expected traffic loads.

The problem can be further broken down by examining the situation as a multiserver loss system. In such a case, the ground rule is that incoming requests will immediately be assigned to a trunk, if such is available; otherwise, the request is rejected without being assigned to a waiting line, since there are none. This type of system can be appreciated more easily by using a graphical representation to illustrate the key aspects involved. The graph in Figure 8-8 provides the information necessary. The designer decides the probability of a lost customer that he or she is willing to accept. This level of acceptance is located on the y axis as labeled. The cumulative traffic intensity, as illustrated above, is determined and the calculated value is found on the x axis. The intersection of these values locates the approximate number of servers needed to satisfy the traffic intensity and no-loss requirements for that scenario.

The model from which this type of system is derived is shown in Figure 8-9. The telephones represent service requests to the central office in the vicinity. Calls for which switching capacity is not available, are provided with a busy signal and the callers must try again later. Those for which there is capacity available are provided to the output trunks, or lines, and these transmission paths are seized for the duration of the calls.

Summary:

Performance monitoring and control systems embedded in an intelligent environment are in the midst of a development and proof-of-concept phase. Many of the ingredients associated with this intelligent environment are still considered leading edge in terms of their current degree of development and sophistication. Nevertheless, these technologies, such as domain understanding, interaction scenarios, multiple database selection, and natural language processing, will constrain the full capabilities of this area until they have been refined.

References

(1)   Ericson, C., L. T. Ericson, and D. Minoli (eds.), *Expert Systems Applications in Integrated Network Management*, Artech House, Inc., Norwood, MA, 1989.

(2)   Feder, B. J., "Frito-Lay's Speedy Data Network," *New York Times*, November, 8, 1990.

(3)   Goyal, S. K., and R. W. Worrest, "Expert System Applications to Network Management," in *Expert System Applications to Telecommunications*, Liebowitz, J.(ed.), John Wiley, New York, 1988, p 1.

(4)   Harris, C. J., and I. White (eds.), *Advances in Command, Control and Communication Systems*, Peter Peregrinus Ltd., London, UK, 1987.

(5)   Henning, W., "Bus Systems," *Sensors and Actuators A*, 25 - 27 (1991), pp 109-113.

(6)   Hoffman, M., "Technology Profile Neural Networks," *Techmonitoring*, SRI International, July 1991.

(7)   Hopple, G.W., *The State of the Art in Decision Support Systems*, QED Information Sciences, Inc., Wellesley, MA, 1988.

(8)   Lemmon, A., *Marvel - A Knowledge-Based Planning System*, GTE Laboratories, Internal Report, 1986.

(9)   Loomis, M. E. S., *Data Management and File Structures*, Prentice-Hall, Englewood Cliffs, NJ, 1989.

(10)   Richardson, J. Jeffrey (ed.), *Artificial Intelligence in Maintenance*, Noyes Publications, Park Ridge, NJ, 1985.

(11)   Singleton, W. T., "Man-Machine Aspects of Command and Control," *Advances in Command, Control and Communication Systems*, Harris, C. J., and I. White (eds.), Peter Peregrinus Ltd., London, UK, 1987.

(12)   Wagner, U., "Reliability and Fault Tolerance of Low-cost Multipoint Sensor Interfaces," *Sensors and Actuators A*, 25-27 (1991), pp 73-78.

(13)   Walker, T. C., and R. K. Miller, *Expert Systems 1990: An Assessment of Technology and Applications*, SEAI Technical Publications, Madison, GA 30650.

(14)   Wallach, R. M., "The E.I.S. State," *Computer Systems News*, March 3, 1990, p 49.

(15)   White, F. E., and J. Llinas, "Data Fusion: the process of C^3I (Command, Control, Communications and Intelligence)," *Defense Electronics* vol22, p 77, June 1990.

(16)   Wilson, G. B., "Some Aspects of Data Fusion," in *Advances in Command, Control and Communication Systems*, Harris, C. J., and I. White (eds.), Peter Peregrinus Ltd., London, England, 1987.

(17)   Richardson, J. J. (ed.), *Artificial Intelligence in Maintenance*, Noyes Publications, Park Ridge, NJ, 1985.

Chapter

9

Neural Networks: Identifying Their Potential

Chapter Highlights:

Neural networks have been around for 100 years or more. Only within the last 10 years, however, have the technology and implementation techniques been available to seriously implement neural network models. Neural network technology mimics its biological antecedent. Investigations of the vertebrate nervous system have revealed architectures and processing methods that originally were simulated on computers, and later were borrowed to develop new approaches to the whole issue of computerized adaptation and learning. Parallel computer systems and distributed computer systems have been the principal areas in which this technology area has progressed. Massively parallel systems incorporate many processing nodes, each of which is its own computer, and is used as a correlate for a neuron. The neuron is the basic communications element of the human body. Distributed systems allow information sharing among the various processing nodes of the system. This chapter surveys the basic knowledge about the neuron and some of the learning, memory, and applications associated with neural networks.

9.1 Neurological Basis of the Subject

The following subsections contain detailed information about the anatomy and physiology of the basic unit of the vertebrate nervous system, the neuron. Any understanding of neural networks as they are implemented in computer systems warrants an equivalent understanding of the underlying basis for these models. Enough is scientifically known about the basic units of the nervous system that their functional and architectural configurations, or, at least, possible configurations, can be used to develop reasonably credible models and simulations from which analyses of and learning about complex systems, such as the brain, can be drawn. More importantly to the network management designer, these models and their subsequent simulations can be used to develop adaptive, learning, monitoring, and control mechanisms for many important network management applications. As will be seen, telecommunications networks can be thought of as the nerve backbone for all purposeful systems devised. These integrative communications links carry the status and actions of the system, thus providing the fabric for learning and adaptation.

9.1.1   The Neuron

Like so many other inventions, the analogy of the human brain as a model for information processing and computer control has been cited once again as a paradigm for the handling of complex events. In particular, the detailed architecture of the brain, i.e., the network of nervous tissue supplying the sensory stimuli to the organism, and the motor responses to the musculature and visceral organs, has come under investigation from engineering circles. In order to understand, therefore, the utility of neural nets as an analogue for complex information handling, we must first understand how and why researchers have chosen neuroanatomy and neurophysiology as paradigms for the applications of computers to human reasoning.   Provided below is a brief outline of the underlying neuroscience of neural structure and activity.

The vertebrate nervous system is composed of a staggering array of specialized cells that relay electrical signals to various points within the organism, e.g., reflex centers, and the brain. These electrical signals are transduced from cells that act in response to very specific stimuli, such as pressure, light, sound, heat, damage to the organism (pain), and gravity. The cells that react to such specific stimuli are called receptors. Other cells transport these electrical signals from the receptors to places in the organism where they are interpreted and acted upon. These specialized transport cells are referred to as neurons and can either facilitate or inhibit the relay of electrical signals depending upon which kind of neuron they are, where they are, or what their function is.

Various estimates have been made about how many neurons each type of organism has, and the human brain is the most popular for such estimating games. Various estimates of between 10 and 500 billion neurons per human brain have been derived, but any such number is somewhat irrelevant, because the issue is what these neurons do and how they do it, not how many there are.

The logical model for the basic nerve component of the vertebrate organism was first proposed by W. S. McCulloch and W. Pitts. [13] This work led to investigations of this and other proposed models intended for the study of nervous tissue functions with special attention paid to the circuit relationships that they established in accomplishing their tasks, among which is information retention. However, the nature of the chemical messages involved, the cytological structure of the neurons, and other biochemical and micromolecular findings about the neurons were not thought to be important to the engineer for an understanding of the neural net architectures and their operational characteristics.

The extension of these investigations into computer modeling and simulation with objectives aimed at human problem-solving has simply been referred to as *neural networks*. It turns out, however, that an in-depth understanding of vertebrate anatomy and physiology is necessary to appreciate how to emulate neural actions. Thus, the replication of neural processes is useful for the scientific world to use in solving real-life problems of importance to telecommunications, for example. Provided below is a more in-depth treatment of the subject of neurology so that the reader can appreciate the rules involved in the solution of such problems.

The basic unit of the vertebrate nervous system is the neuron, or nerve cell. It consists of a cell body, *afferent* structures (signal reception fibers, called dendrites), and an *efferent* structure (a signal transmission fiber, called an axon). All neural processes in the organism call for some form of stimuli, many times occurring externally, that are picked up by certain structures such as the eye or ear via special *receptors* suited for these stimuli. Other stimuli are endogenous, that is occurring internally, and may be associated with certain memory functions or visceral sensations.

A *receptor*, which is a specialized sensor specific to a particular type of stimulus, such as impressions of pain, will react to that stimulus by passing an electrical signal in the form of a depolarization wave to the first neuron with which it comes into contact. From there, the signals are passed as *action potentials* to succeeding neurons until these signals are received in the brain or on efferent neurons that generate motor responses. Different stimuli are coded according to the number and/or frequency of action potentials generated. These coding schemes convey certain meanings to the brain according to their organization. The coding of neural data, by which there is a distinction among stimuli, is accomplished by any number of methods, that are outlined below:

(1)   Frequency of Pulses

The continued stimulus to a receptor will cause an increase in the pulse rate output by that neuron. There is also a perceived increase in the frequency of pulses received by the central nervous system (CNS) as the stimulus intensity increases, that results in the recruitment of more and more neurons.

(2)   Spatial or Unit Discrimination

The CNS has little difficulty in distinguishing among various stimuli, such as between light and sound, or between pressure on the buttocks as the organism sits down and a sharp pinprick to the hand. It is a characteristic of the CNS that it has the ability to locate spatially the sources of different stimuli.

(3)   Time-Rate Variations

The differential firing patterns of a neuron provide clues as to whether the neuron is being stimulated more or less. The timing of the firings and their intensity are the clues for this phenomenon.

(4)   Crossmodal Correlation

The combination of neural data coming from different modalities provides added confidence, in the case of attention queues, or added appreciation, in the case of esthetic queues.

(5)   Experience and Memory

Each new neural exposure adds to or detracts from previous stored experiences that have been retained by the organism. Thus, informational value is constantly shifting based upon overlay upon overlay of each new exposure to the world. Earlier experiences tend to provide the initial frame of reference, and each succeeding experience is viewed in context with that starting reference.

The energy stimulus at the receptor, whether it be sound, light, etc., causes an electrical signal, or depolarization, to be generated by the receptor. This depolarization is the input side of the neural process. The electrical wave in turn causes a subsequent deposition of transmitter chemicals to be secreted through the *presynaptic membrane*, i.e., the membrane of the receptor. These transmitter chemicals then migrate across the synaptic space to the membranes of the dendrites of the first afferent neuron.

The dendrite then responds to these chemicals by creating yet another depolarization within the *postsynaptic membrane* and an electrical wave begins to travel toward the cell body. This, therefore, is the output side of the neural process. This chemical interface point is known as the terminal *bouton* or termination end of the receptor. The chemicals rapidly cross an interstitial space known as the synapse, and interact with the dendritic membrane of the afferent neuron in contact with that receptor.

Not all neural activity is excitatory, or a source of excitation. Some neuronal interactions are inhibitory, which means that the secretions that pass across the synaptic junction between nerve cells actually inhibit nerve firings. Logically, it would appear that there is no need for a neural network arrangement to have *inhibitory neurons* unless there were a possibility that an excitatory reaction could occur. Thus, it seems that there is no need for inhibitory neurons by themselves.

If the neuron is excitatory, the electrical waves traveling from the dendrites reach the cell body, and in particular an area known as the axon hillock, where the waves coming from the various dendrites, if of sufficient strength, cause an electrical spike to be initiated. This electrical spike is the *action potential*, and is self-sustaining all along the entire length of the *axon* of the *neuron*. In the afferent neuron this action potential travels down the axon to the next synapse or junction between neurons, and the process is repeated. This process can be repeated up to about 1000 times per second in some cases, but much less often for the average neuron.

The rate of neuron "firing" or repetition of signaling, called the *action potential*, is indicative of the intensity of the stimulus. The action potential is governed by what is called the *all-or-nothing principle*. This means that there must be a sufficient number of secretions by the receptors or preceding neuronal axons to create a similarly sufficient energy depolarization at the axon hillock of the neuron to initiate the action potential. If there is not a sufficient depolarization at the axon hillock, the action potential will not be initiated.

The action potential is actually a large, sharp electrical wave created by a sufficient number of smaller depolarizations at the nerve's dendritic, or front, end. These dendritic depolarizations travel from the receiving sites up through the nerve branches to a point near the cell body where the action potential is triggered if there are a sufficient aggregate of electrical depolarizations present. Figure 9-1 shows the basic arrangement of the sensory, or incoming signal, transmission system.

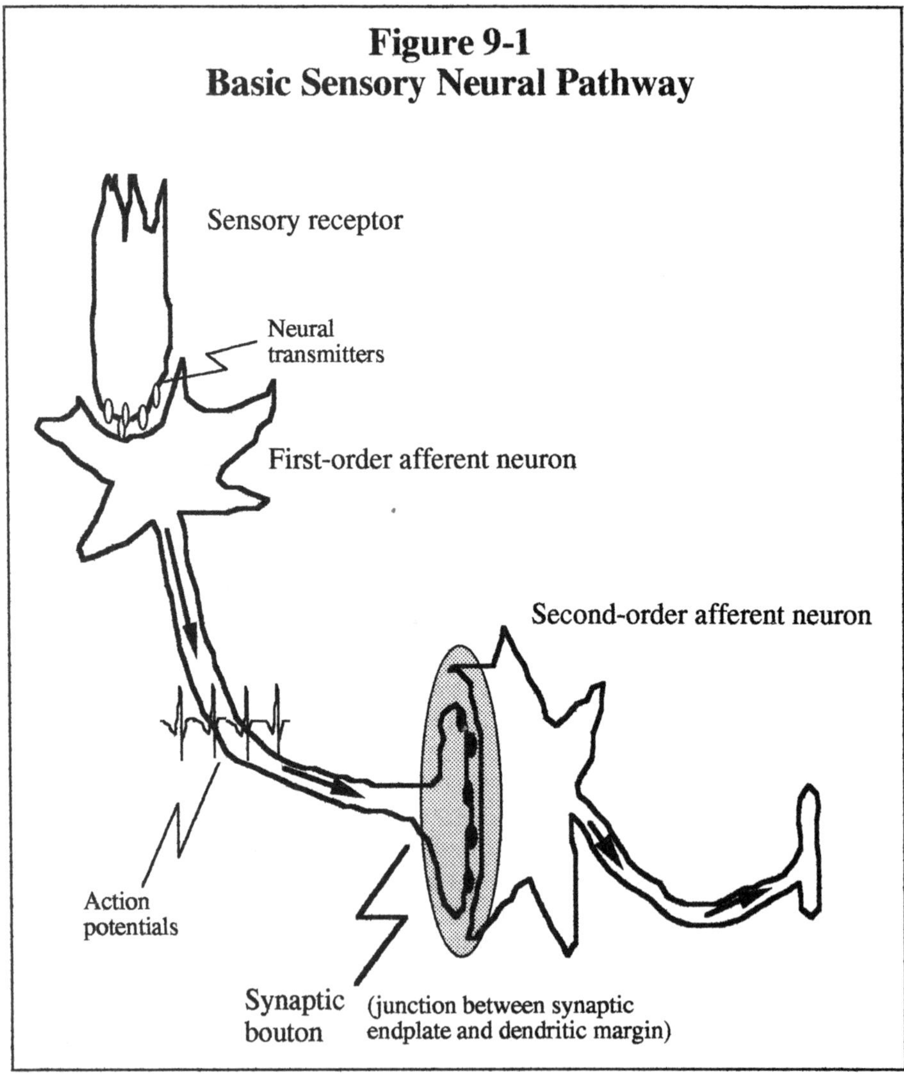

9.1.2   Neural Circuitry and Fusion

There are several mechanisms, however, which assist the sensory mechanism and/or memory to sustain a more intense reflection of the incoming or reverberating signal. These mechanisms are known as *recruitment, differential pathway transmission, neural divergence* (amplification), and *neural concentration.*

With the first mechanism, recruitment, more neurons are called into operation as the intensity of the signal increases since different neurons have different thresholds of activation or action potential generation. Thus, a receptor may generate a signal in a first-order afferent neuron, whose intensity may at first only cause a series of action potentials in one of several interfacing second-order, or succeeding, afferent neurons. As the intensity of the signal in the first-order neuron increases, an additional second-order neuron may begin firing, or sending out an action potential along its axon. Now we have two neurons carrying signals caused by the receptor, the difference being an increase in intensity of the receptor's signal.

The second mechanism, differential pathway transmission, can be thought of as analogous to differential transmission of a signal along two or more electrical wires. The differential signal path effectively increases the signal-to-noise ratio of the transmitted signal. In the brainstem and other places in the brain, there are several examples of dual or differential pathways. These paths, in some cases, are thought to be recruited for intense signals and in other cases are thought to enhance the signal threshold for higher-level centers in the midbrain.

Neural divergence, or amplification, is achieved through the first-order neuron making synaptic contact with several succeeding neurons on their way to processing in the brain. These following neurons are called *interneurons*, because they perform a relaying function only. The overall effect of neural divergence is an amplification of the original signal that caused the action potential. Since this is a one-way signal process, the net effect is that of a rectifier-amplifier.

Neural concentration is just the opposite from that effect described above. It is achieved when several neurons terminate on one following neuron. The effect assures that the *action potential* reaches its intended point of usage, by forcing a redundancy upon the the following neuron to guarantee signal relay.

9.1.3   Representations of Neural Models

The logical representation of neural modeling, as depicted in Figure 9-2, actually illustrates that different events can be described by the same set of neural inputs. [2] These events can be visual, auditory, olfactory, pressure, or functional. Thus, the neural network is only specialized to the extent that it is employed to service a particular type of modality. In the case of technology, events may be associated with the operations of a network management system. In the figure, sensory events are presented to the first set of neurons, in the perceptron (multi-level array of neurons), as it was named by Rosenblatt [15]. If the set differs slightly in terms of the number or type of neurons that are firing, this difference can be interpreted as a different result.

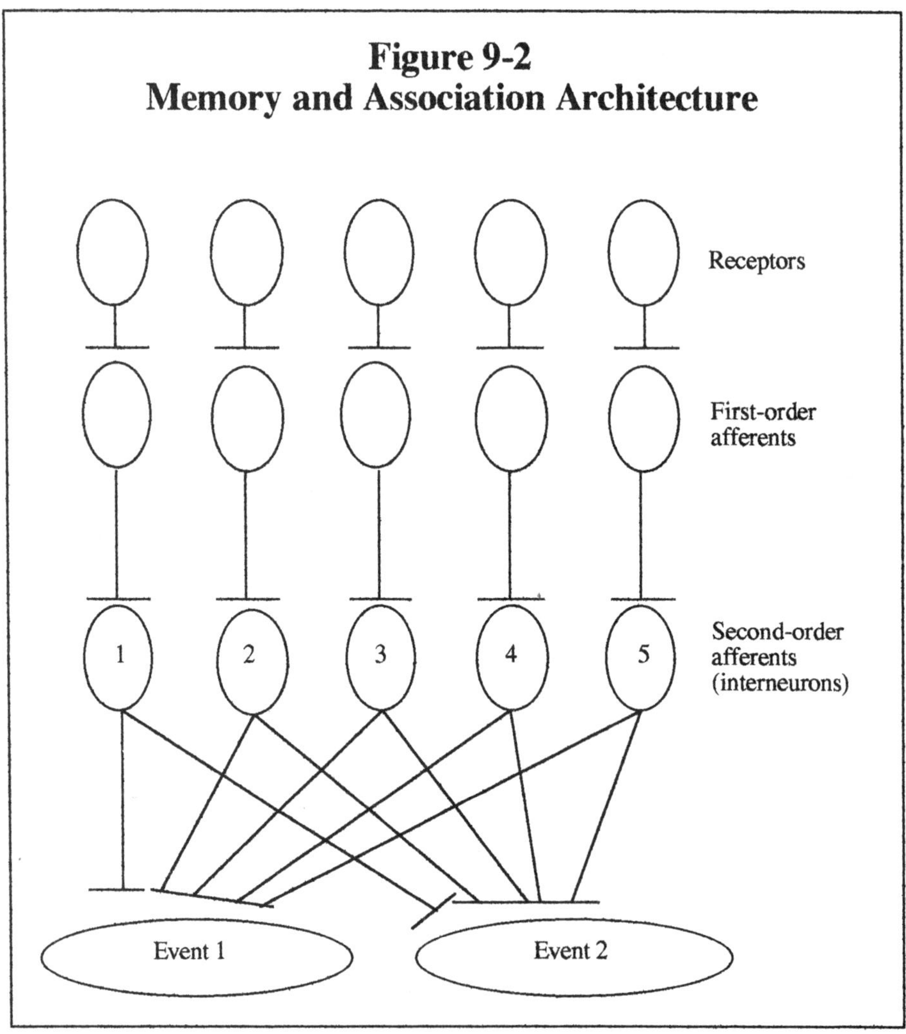

An example of such a difference could be easily imagined by thinking of the difference between the letter P and the letter R. The leg on the R could be detected as a result of additional sensory input from other receptors. These additional receptor inputs will then be integrated and interpreted in the total integrated result as the letter R. The outputs of the perceptron, Event 1 and Event 2, may be different only in the value of one input (receptor activation). This slight difference may be enough to distinguish between the letters R and P.

Each of the incoming signal points, or receptors, has a bias associated with it. This bias is impressed upon the computer-simulated version of the neurons, with the effect of increasing the frequency of the firing rate of the neurons. This bias now becomes a factor in each neuron's influence on the overall reception process, and possibly subsequent influence upon the event identification with which it is associated.

The question has arisen as to how the neural network can achieve consistent results, store such results, and associate these results with later inputs. These latter two issues will be addressed in subsequent sections. It is widely believed that neurons are basically unreliable in that they tend to misfire, fail to respond to firing thresholds, and fire at variable rates even though the stimulus is consistent. If such elements are inherently unreliable, then how are they to maintain a consistent set of actions over time?

The answer is that they probably have a considerable amount of redundancy embedded in the various circuits for each event to be stored away. Neural networks appear to have some redundancy of fibers carrying messages about the cortex. The notion that each of these groups of fibers used to collect, sort, or interpret an event use the polling technique where the majority determines the outcome has been suggested. Thus, reliability and retention are achieved.

This approach is best appreciated by the classic example of the redundant system elements where the lack of reliability in one is negated by having two or three such elements in parallel in the system. Thus, such an arrangement will guarantee, to a very high level of probability, that the signal will get through the network. This arrangement is illustrated in Figure 9-3.

Traditionally, reliability has been a measure of consistency, or consistent results. As a measurement parameter, it is calculated in several ways, such as by using the history of the equipment to make judgments, making use of parts counts to derive a complexity factor that can be converted into a reliability number, or evaluating risk factors in the technology that can also be translated into a reliability value. [7] Reliability implies the ability of a system or component to provide full service capability between failures. For this reason reliability is somewhat analogous to mean time between failure. Many times, reliability is the single most valuable measure for determining whether a certain system configuration is sufficient for operational needs or whether additional or substitute equipment is required. It can, more than almost anything else, affect purchasing decisions for a system.

If we have a system that has a certain reliability value, such as 75 percent, or 0.75, we know that the configuration of the represented equipment functionality will demonstrate its designed functional response about 75 percent of the time. Of course, there is no assurance that the response is appropriate; this would involve a different design parameter, namely validity. If we are satisfied with the validity of the response but want a higher reliability value, something has to be done. We could look for another piece of equipment that meets the required reliability threshold, lower our expectations to fit the reliability promised, or use the current equipment in a redundant configuration.

The use of a redundant configuration also requires that additional equipment be employed to sense the failure in one piece of equipment so that the other may be committed to service. If we choose a redundant configuration where the failure of one system will trigger the use of another, what might we expect in terms of overall reliability? The answer can be reached by employing the following equation:

$$TSR = 1 - (1 - SAR)^2$$

where: TSR = total system reliability
            SAR = stand-alone reliability

For example, two units each having an individual reliability of 0.75, or 75%, would yield the following composite probability, if operated in parallel.

$$\begin{aligned} TSR\ &= 1 - (1 - 0.75)^2 \\ &= 1 - 0.0625 \\ &= 0.9375 \end{aligned}$$

Thus, for the example cited above, if two units are to be used with a reliability of 0.75 each, the combined reliability will be 0.9375. This is the basis of fault-tolerant systems design. An illustration of such a problem is provided in Figure 9-3.

Another approach can be developed from a more data-oriented view of image processing. In image processing a pixel stands for a picture element within the field of the image. Each pixel can be thought of as a point within that field of view where information can be displayed. Such information can be light intensity and color. An array of these pixels, taken together, portrays an image that incorporates these elements with their color and intensity in order to make up an image. In the case of a language interpreter, a scan of the letters P and R could result in extra signals being received to detect the difference between the two letters and a second step taken to interpret what these differences mean. If a row of pixels at a time is processed, the resulting pattern derived from this action yields a difference between the letters R and P, as illustrated in Figure 9-4.

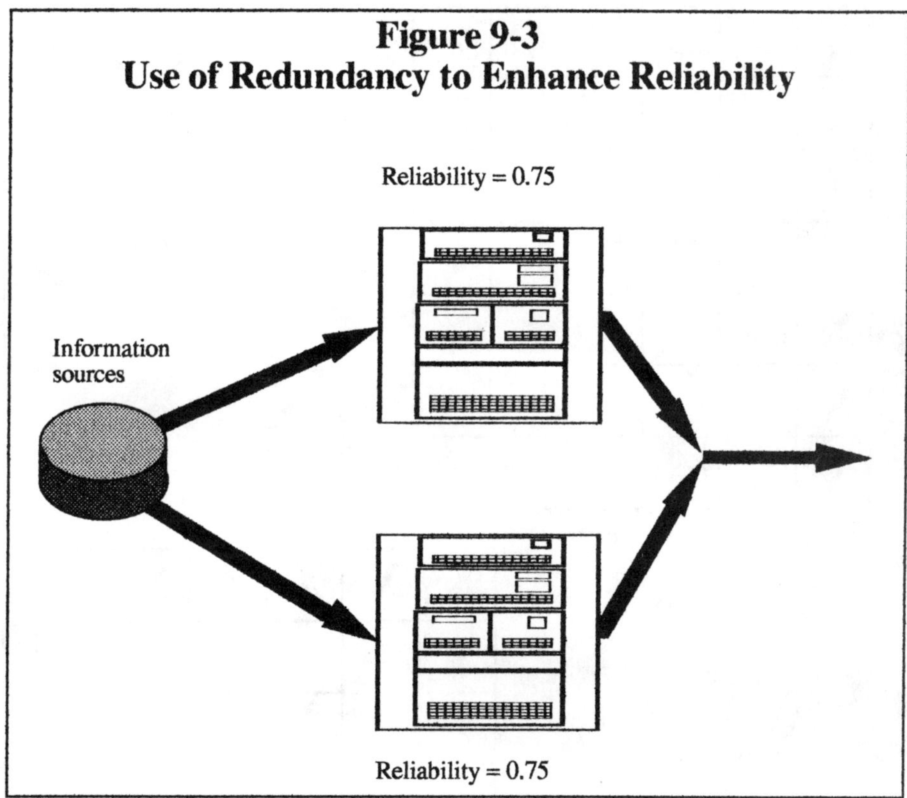

Figure 9-3
Use of Redundancy to Enhance Reliability

To this point, the process can be equated to machine vision, but neural networks carry the acquisition and interpretation activity to a different level. For instance, a neural network version of the letter interpretation problem would acquire the entire image and end up with a different output than another acquisition that contained a different letter. The interesting aspect of neural nets is that they are not geared toward a specific sensory modality. They are general in the sense that presumably any neural net can be used to process any sensory mode. Additionally, the

Figure 9-4
Neural Network Approach to Pattern Recognition

X	X	X	X	X	
X					X
X					X
X					X
X	X	X	X	X	
X					
X					
X					

X	X	X	X	X	
X					X
X					X
X					X
X	X	X	X	X	
X			X		
X				X	
X					X

outputs are not specifically coded to reflect any particular mode of sense. This particular coded result can now be used as interpreted, or retrieved as necessary as a memory element for reference and use.

9.1.4   Uses of Neural Networks

Recent neural networks have been used not only to match prestored symbols with similar symbols detected by the system's sensory input apparatus but also to acquire new symbols that were not prestored but associated with certain new situations. In earlier times when optical character recognition and machine vision were in their infancy, these neural-network-like functions were performed by machines that were special-purpose computers and could recognize a limited set of handwritten symbols using the concept of random inputs to the processing apparatus from the sensor arrays. Early attempts to perform neural-like analyses of such topics as handwriting and vocalizations were approached in the same ways as standard data processing problems. These approaches clearly did not work.

A more detailed explanation of neural networks is illustrated in the basic perceptron model provided above. In this model, the neural network is composed of a receptor field of S elements, to which is attached a set of afferent neurons called A elements. Each A element has some number of receptors attached, but not all receptors are attached to each A element. A elements, in turn, are attached to an R element in a second layer of afferent neurons, such that the degree of stimulation on the R element can be made variable depending upon the number of A elements providing the stimulation to each R element. Some imagination may be required in order to verify for one's self that this configuration will, in fact, provide the imprint desired.

However, no matter which explanation is most comfortable for the reader, the point is that incoming events can be captured, distinguished from other events, compared to new events, and indications provided to a decision layer as to the results of those comparisons. In theory, very slight differences in behavior can be detected using this approach and the technology derived therefrom. This is the primary advantage of neural networks; i.e., they can detect very slight differences and still associate such differences with an identifiable object.

Once the signal reaches places where it can be acted upon or compared and associated, two additional processes may be involved in the "imprinting," or memory function, these being habituation and sensitization. A full description of these processes is beyond the scope of this book. However, *habituation* is roughly analogous to a dulling of the senses, while *sensitization* is analogous to a more intense sensation. Beyond this, certain sensory, memory, or relational events can resurrect previous experiences or assist in framing associations that can be used by the organism using habituation and sensitization mechanisms. The analogy to be used in computer technology is that of the reverberating circuit or elements. In the computer world this is also accomplished by defining a series of elements similar to the arrays of elements used in parallel processors for the storage of events that are to be retained.

Putting these concepts together, we might visualize a situation where incoming sensory data must be recognized, stored, and recalled for later comparison. An input to a bank of elements may temporarily imprint an event upon that group of elements. Another input may imprint yet a slightly different event upon that group, even though some of the same elements were used as in the first imprint, but slightly different elements were used in the second imprint. If any one of several choices of previous groupings could be recalled and compared to some later stimulus with its established imprint on an array of elements, then a favorable comparison could be declared a match, and appropriate action is taken or duly noted; an unfavorable comparison causes the comparison process to continue until a favorable comparison is made or until the comparison possibilities are exhausted.

The imprint process is one that is of consuming interest to the researcher and/or clinician. Imprints are thought to be provided through reverberation circuits. Such circuits are essentially a series of neurons that form a closed loop, as illustrated in Figure 9-5. These closed loops hold the signal for later recall as needed by the organism. This is the basic paradigm for memory and its psychological corollary, learning. The event is imprinted upon the reverberatory circuit and when needed, is read out to the processing or association processes for comparison and use.

9.1.5   Correlates to Network Management

In the area of network management, neural networks have many uses. A neural network previously "trained" in recognizing certain event combinations, or operating signatures, might take action that would lead to an equipment reconfiguration of that system. Such a reconfiguration might be instituted to accommodate an increased system loading, or to create an alarm that otherwise might go unnoticed for some time, or, possibly, to initiate an unscheduled maintenance action. Signal processing is one such area where neural networks can be used effectively to analyze signals for their use in decision making.

The larger the number of monitoring points, the more these points begin to appear like a complex signal. Thus, their decomposition is more directly handled by neural processes. For network management, the possibilities of incoming and potentially competing signals are staggering.

The idea that a seemingly random set of monitor points, whose outputs identify the existence of an event requiring attention, and whose output can be provided to a predetermined set of receivers, is made to order for network management situations.

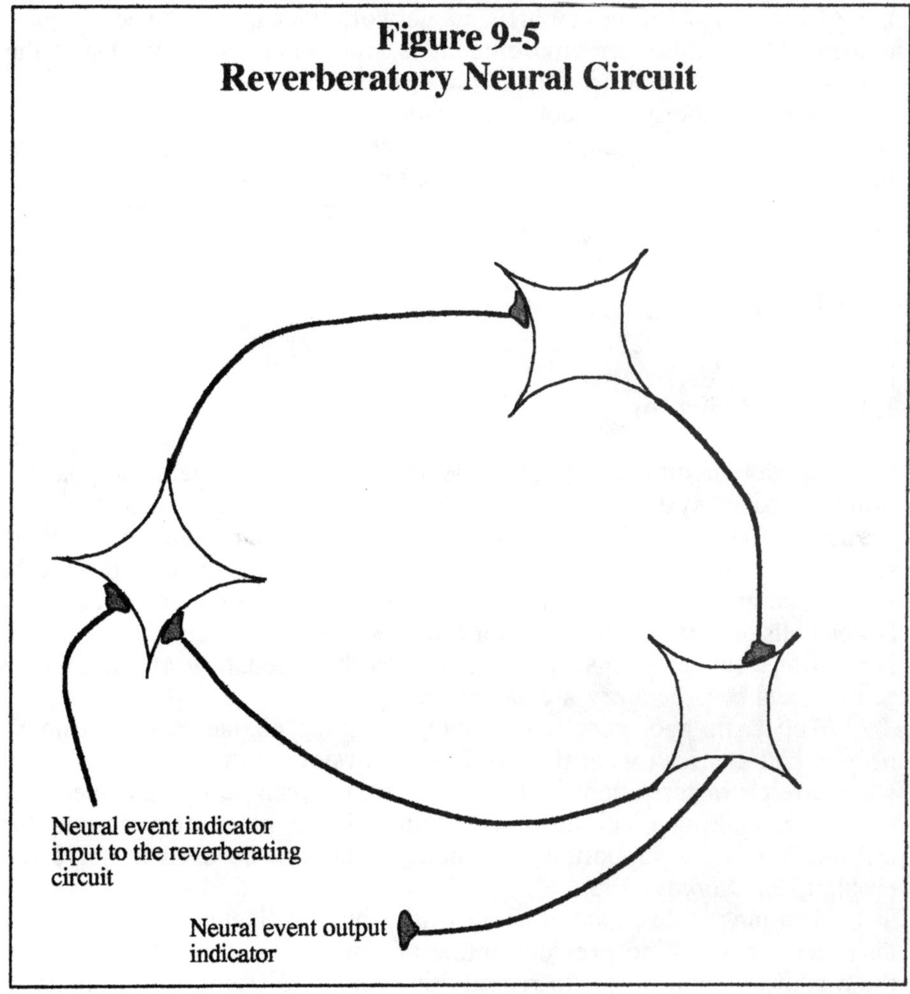

Figure 9-5
Reverberatory Neural Circuit

Typically, network management environments are not made for nice, neat monitoring. These environments are complex and are built and upgraded for operational necessities, not management. Little prior thought is given to the issue of how this functionality is to be monitored in an integrated fashion. It would be valuable if monitored connections could be randomly connected to any one of several integrating points and their output relayed to a management station for automated analysis and decision support.

Neural networks are also being considered for many noncommunications tasks as well, such as reading, handwriting analysis, and the observation of visual scenes for certain decision parameters or queues. The use of neural nets in network management seems quite appropriate, because the environment is somewhat constrained and the results can be measured so much more directly.

One of the original motivations for the investigation of the nervous system was to understand better the subject of learning. Learning and its close correlate, memory, will be discussed in the next section.

9.2 Learning and Memory

9.2.1   Learning

Learning has often been thought to be an elusive and unattainable goal for computer-based systems until recent times. The realization of this elusive goal, if realized, is likely to come as a result of the study of nervous systems of various vertebrate and invertebrate animals, such as the leech, sea urchin, and crayfish. These animal models have simpler systems and larger cells in some cases that lend themselves to the detailed study of interactions between sensory and motor tracks, excitatory and inhibitory actions, and reverberatory architectures.

Ten years ago, very few people, including engineers and business people, had any idea what the word neurophysiology meant. Neural nets, which are characterizations of neural processing, can be implemented in a computer, making disciplines like neuroscience more of a source for information and a support for technology that can be used to solve the problems of everyday life.

Learning is that state of being where the organism becomes aware of an event for which no previous frame of reference exists. At this point, the organism decides how to regard this event, and stores that decision as information. Subsequent events will either confirm or deny some previously established frame of reference. If none exists, the organism will store this version and recall it for association or comparison as circumstances dictate later.

Learning begins with perception. Perceptions are the result of sensory inputs that invade the organism many times per second. These perceptions propagate to the brain where they can be detected using electroencephalography (EEG). Perception is an aggregate of many action potentials reaching the brain at any point in time. The question is how these perceptions become durable, or stored, for later retrieval. The retrieval of these stored perceptions, or engrams, is the reverse of the imprint process and can be triggered on demand in most cases by the human organism. The search for this retrieval mechanism has been without success so far.

Various animal models, as well as human lesion studies, have shown that certain areas of the brain are specialized for the learning of particular tasks. For example, visual learning can be impaired by infarctions (hemorrhaging) and ablations (physical damage) to the visual cortex in the occipital area (rear) of the head. Similarly, the parietal (top, rear of center) area of the brain is responsible for our ability to do tasks, such as button shirts. However, even if a small portion of that cortex is left intact in the rat, the animal is still capable of normal learning. This leads to any of several conclusions, all of which imply that learning can be transferred to some degree, and that generalized learning capacity protects the organism, or system, from widespread incapacities.

Also, monkeys trained with different numbers of trials to perform a particular task show a stronger tendency to continue performing that task that required the larger numbers of training trials. This suggests that, possibly, the longer the learning period, the more widespread the learned experience becomes and the less likely that experience is to disappear, or be forgotten. In humans with brain damage afflictions, their abilities to relearn are not lost, depending on the extent of damage and the age of the individuals involved. Some notable exceptions appear in humans, however.

Apparently, because of our tendency toward specialization, as we develop after birth to maturity, our brain areas become more and more resistant to the redistribution of sensory and motor functional control to other areas of the brain as a result of some type of insult to those specialized areas. For instance, the age of 12 is about the point at which a person can relearn language if some type of neurological insult blocks that area of the brain (Wernicke's area) responsible for language.

The lessons here for man-made neural networks are only partially clear. First, neural networks should be plastic, or flexible, enough to accommodate changing situations in the operational environment. Second, learning should capable of restarting from some beginning point. Neural networks implemented in telecommunications systems should, therefore, be general enough that they can fail gracefully and have the ability to relearn depending upon the damage to their internal systems. Also, the longer the experiences, the more pervasive these learned experiences become, and the less likely these experiences are to be forgotten.

9.2.2   Memory

Memory is the psychological reference to the mechanism that allows learning to be employed over and over again.   Physiologically, interpretations and responses that work or don't work for the organism are stored in the brain by some device or devices that are thought to be related to different kinds of feedback mechanisms, and therefore, create reverberating circuits within the neural matrix of the brain.   This neurological feedback can be thought of as a version of the feedback paradigm from cybernetics and control theory.   This paradigm is illustrated in Figure 9-6.

The neuron has some mechanism, yet to be identified, that allows its message to be regenerated indefinitely until used or until that message disappears as a result of neural disintegration. Such a circumstance might be analogous to forgetting.   In any case, the neural feedback may be supported by the mechanism of reverberation, as was illustrated in Figure 9-5.   In an engineering sense, the feedback can be achieved within file structures and database systems.

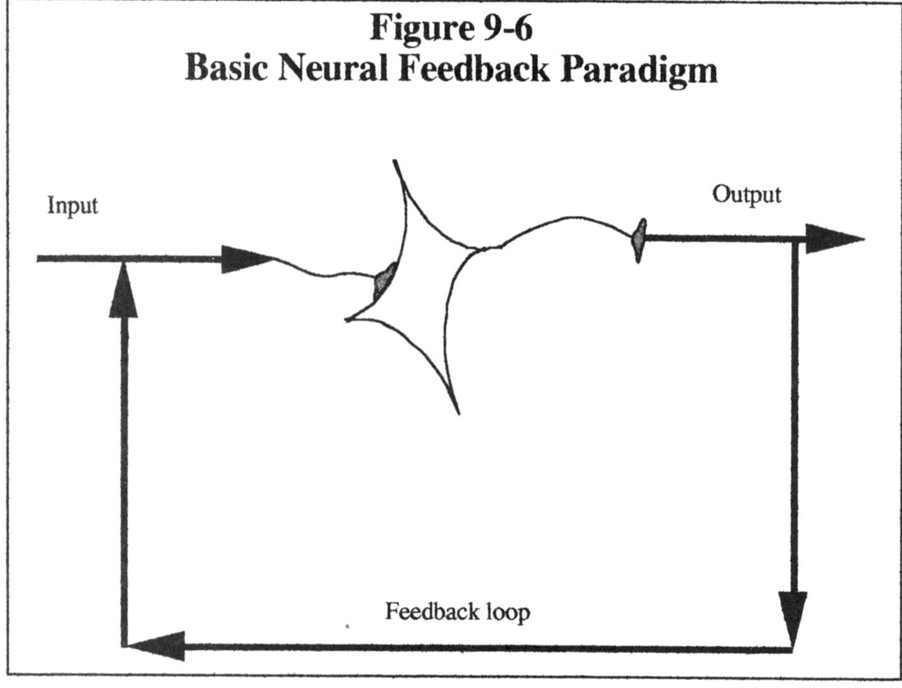

**Figure 9-6
Basic Neural Feedback Paradigm**

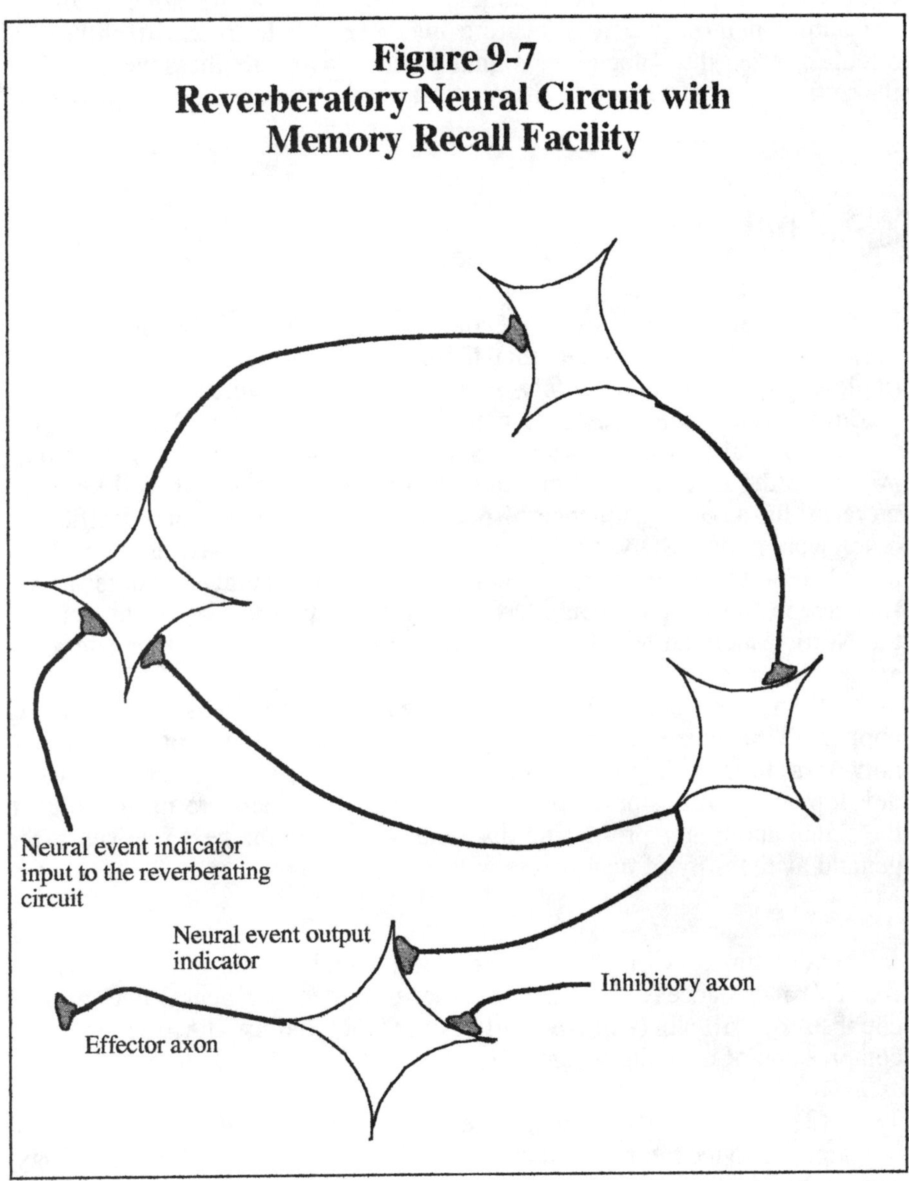

Memory recall is achieved by providing or eliminating some stimulus to the reverberating circuit that is recirculating the retained signal. Figure 9-5 showed that a signal can be captured and recirculated within a reverberating circuit. The recall of that signal must now be determined. It can be retrieved by a means similar to that illustrated in Figure 9-7. As long as the inhibitory neuron sends a "negative" signal to the output neuron, the recirculating message will be retained within the circuit. When that "negative" signal is shut down, the message will flow through the output neuron to the next stage of action or interpretation.

9.3 Applications

Professor Bernard Widrow is recognized as being the father of neural networks, and he has stated that there is an enormous potential for the implementation of neural networks in the area of telecommunications. Additionally, he believes that neural nets will make ISDN (intelligent services digital network) possible as a viable network service. This is due, in part, to the fact that the limitations of the ISDN bandwidth will be made more useful as a consequence of neural net processing of data traffic. In other words, the ISDN will become more efficiently used as a result of neural network monitoring and control. Clearly, neural networks, as we shall see in this chapter, could have a significant part to play in the entry of the North American telephone companies into the enhanced home services arena.

If anything can help loosen the restrictive bandwidths of the existing copper plant currently burdening these telephone companies, neural networks hold the greatest promise. Those in the neural network development and business areas believe that neural nets are most useful in the signal and image processing disciplines. Within the next five years, the general availability of neural nets will:

(1)   make video teleconferencing and high-definition television (HDTV) more accessible and less expensive, because neural nets will be capable of filtering noise and expanding bandwidth through the compression of data and images.

(2)   make natural language, both written and verbalized, and interactive services more available through the maturity of this technology, and its use in various telecommunications applications, such as operator services.

We will now explore some of these applications.

9.3.1 Video Processing

Several research facilities have been working to utilize neural nets as a tool to enhance video signals and to make them more accessible for transport across narrowband transmission media. Work is being conducted, for example, at the David Sarnoff Research Center to develop techniques to understand and filter out noise from transmission signals. The signals of choice, initially, have been composite video signals that are corrupted with AM impulse noise. The method being explored currently is that of utilizing a parallel computer to implement three back-propagating networks that filter, detect, and replace random noise spots, resulting from AM impulse noise, with appropriate video information.

9.3.2 Language Processing and Interpretation

Optical character recognition, at some reasonable rate of accuracy, has been an objective in many areas for years. Probably the best, and most easily understood example, is that of mail routing, and all that this involves. The mail system requires certain specific tasks, most of which involve the understanding and interpretation of addresses. Some of these tasks are routing based upon addressee unknown, actual delivery to the correct and intended address, returned mail because addressee moved, and misaddressed mail.

9.4 Computer Analogs of Neural Networks

Take, for example the basic design of the standard digital computer of today as illustrated in Figure 9-8. This device shows an overall control feature that orchestrates each of the other main functions of the computer, these being the input, output, arithmetic, and memory units. The brain analogues are the sensory, and motor/visceral nervous systems, association, and the memory areas of the cortex. The brain, as yet, has not revealed a central control area or even function for that matter. The basic precept upon which neural networks are based, and have thus far been demonstrated, is that the neural network has no control function as such. The basic analogue for a neural network is illustrated in Figure 9-9.

Here, the network has no apparent control or association areas. These functions are performed by the same connections that are storing the sensory information acquired. The theory is that sensory information constantly flows into the neural network and is imprinted upon some subset

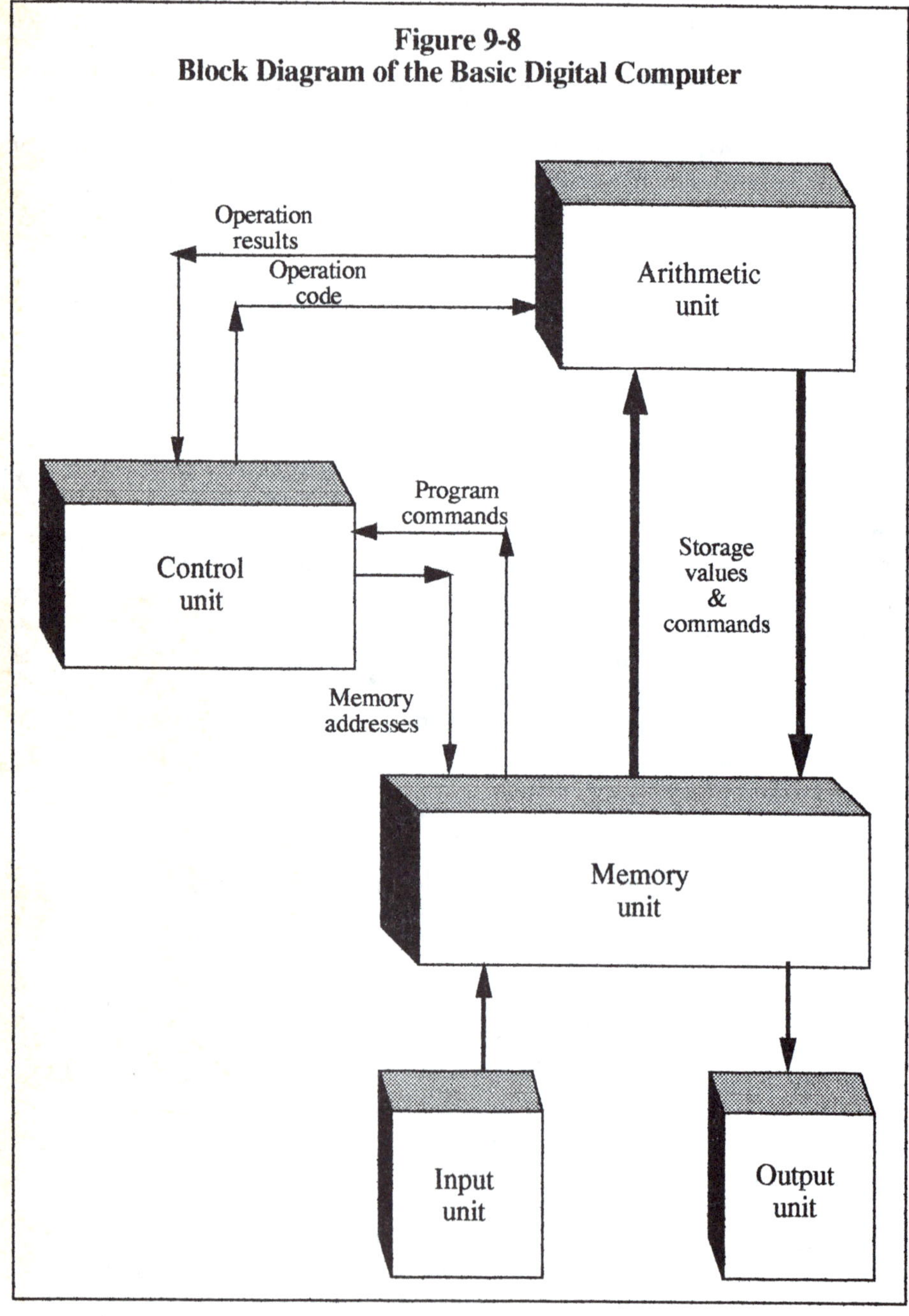

**Figure 9-8
Block Diagram of the Basic Digital Computer**

of the network. The recurrence of certain patterns is more lasting than others, possibly because of their repetition, and the number of redundant fibers carrying the signals produced.

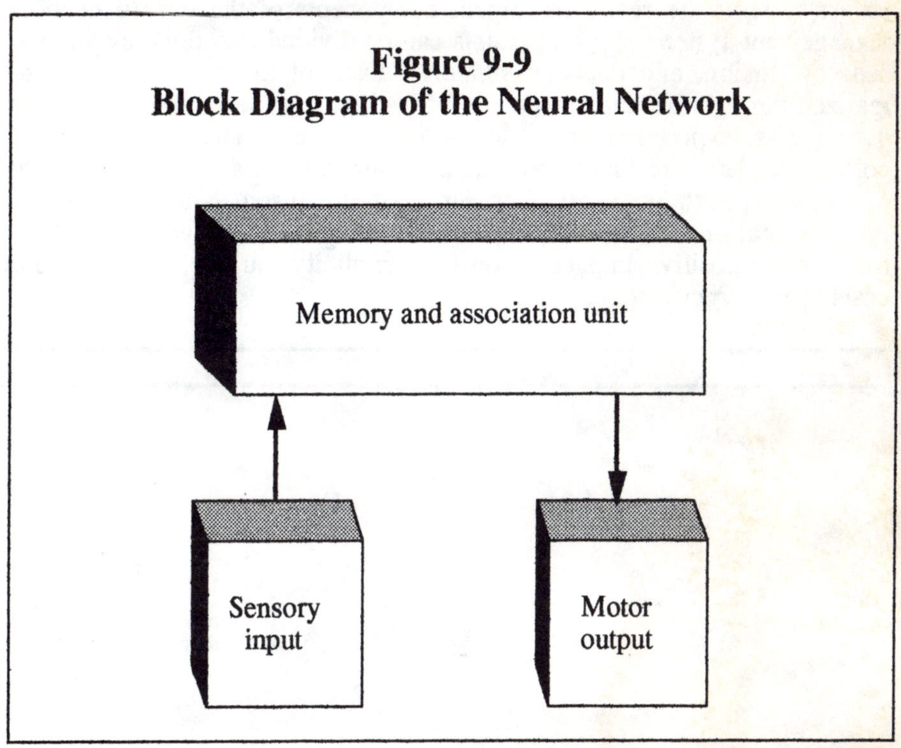

Figure 9-9
Block Diagram of the Neural Network

Summary:

Smart systems are really intelligent components of the overall network management system. Smart systems can be divided into three major sets: sensory, fusion, and decision support. Each of these, in turn, can be realized through the implementation of leading edge technologies, such as neural nets, expert systems, and robotic features. The system sensors collect the data, the fusion process integrates the data collected, and the decision support process makes sense of the fused data. The proper implementation of any of these features into a network can have tremendous positive impacts upon the simplicity, quality, and operating costs of the target system.

References

(1)	Bertuol, B., "Sensors as Components for Automotive Systems," *Sensors and Actuators A*, 25-27 (1991), pp 95-102.

(2)	Ericson, C., L. T. Ericson, and D. Minoli (eds.), *Expert Systems Applications in Integrated Network Management*, Artech House, Inc., Norwood, MA, 1989.

(3)	Feder, B. J., "Frito-Lay's Speedy Data Network," *New York Times*, November, 8, 1990.

(4)	Goyal, S. K., and R. W. Worrest, "Expert System Applications to Network Management," in *Expert System Applications to Telecommunications*, Liebowitz, J.(ed.), John Wiley, New York, 1988, p 1.

(5)	Harris, C. J., and I. White (eds.), *Advances in Command, Control and Communication Systems*, Peter Peregrinus Ltd., London, UK, 1987.

(6)	Henning, W., "Bus Systems," *Sensors and Actuators A*, 25-27 (1991), pp 109-113.

(7)	Hoffman, M., "Technology Profile Neural Networks," *Techmonitoring*, SRI International, July 1991.

(8)	Hopple, G.W., *The State of the Art in Decision Support Systems*, QED Information Sciences, Inc., Wellesley, MA, 1988.

(9)	Lemmon, A., *Marvel - A Knowledge-Based Planning System*, GTE Laboratories, Internal Report, 1986.

(10)	Loomis, M. E. S., *Data Management and File Structures*, Prentice Hall, Englewood Cliffs, NJ, 1989.

(11)	Singleton, W. T., "Man-Machine Aspects of Command and Control," *Advances in Command, Control and Communication Systems*, Harris, C. J., and I. White (eds.), Peter Peregrinus Ltd., London, UK, 1987.

(12)	Wagner, U., "Reliability and Fault Tolerance of Low-cost Multipoint Sensor Interfaces," *Sensors and Actuators A*, 25-27 (1991), pp 73-78.

(13)	Walker, T. C., and R. K. Miller, *Expert Systems 1990: An Assessment of Technology and Applications*, SEAI Technical Publications, Madison, GA 30650.

(14)	Wallach, R. M., "The E.I.S. State", *Computer Systems News*, March 3, 1990, p 49.

(15)	White, F. E., and J. Llinas, "Data Fusion: the process of C^3I (Command, Control, Communications and Intelligence)," *Defense Electronics* vol 22, p 77, June 1990.

(16)   Wilson, G. B., "Some Aspects of Data Fusion," in *Advances in Command, Control and Communication Systems*, Harris, C. J., and I. White (eds.), Peter Peregrinus Ltd., London, England, 1987.

Chapter

10

Neural Networks: Design and Usage

Chapter Highlights:

The design and architectural issues associated with neural networks are the key to their utility in network management environments. Neural networks are presently enjoying some initial successes in their implementation in certain limited applications. But their continued utility and expansion into other more complex situations and environments depends upon the rate of technology developments that will support neural network system implementations. Some of these developments include software developments tools and communications protocols. In addition, their use also depends upon their ability to be integrated into the processing environments of more complicated systems. The long-term picture for neural networks appears to be one of close coupling between neural networks and other more traditional processing approaches. Network management and, in particular, signal processing and machine vision applications, are particularly suited for neural network analyses and resolution.

10.1   Design Issues

10.1.1   Basic Features

Neural networks, whether implemented in laboratory or commercial applications settings, have some special design features in common. The basic elements of neural networks, at a level of more detail than that presented in Chapter 9, are presented here. As previously mentioned, however, neural networks basically consist of multiple system inputs, aggregate or integrated outputs, summing nodes, and weighting functions.

The general architecture of neural networks, like that of their anatomical animal correlates, is organized into layers. These layers are referred to as the *input layer*, *output layer*, and the *hidden layer*. The hidden layer is actually an intermediate layer between the two outside layers, and might be viewed as an interpretation layer. Each layer consists of one or more, usually many more, processing elements, or nodes, that form the basic physical and functional structure of the neural network. Figure 10-1 shows the basic neural network features as they relate to one of these nodes. Output patterns generated are used to feed back information to the input weights associated with each input parameter. [1]

Thus, weighting factors are modified on the basis of desired, or targeted, results that correct actual results, and drive them toward the target objectives. Additionally, each node may, as depicted, typically receive more than one weighted input from external sources. As the reader will see later, intermediate processing at nodes within the hidden layer of the neural network fabric, also allows two or more inputs to each node. In Figure 10-1, X_n are the inputs to each node, and W_n are the weighting factors.

The anatomical correlates of neurons, or nerves, are referred to as neurodes, or nodes, in an engineering setting. In Figure 10-1 each neurode is identified by its value of N, and each receives the weighted inputs of one or more sources, where the weighting factors, W_n, are applied along the connection paths between the inputs, X_n, and the nodes. These paths are represented as directed arrows. Ordinarily, the weighting factors accompany each input and there is normally one weighting factor associated with each input parameter. N calculates the average value of the summed multiplications of the inputs and the weighting factors for each connection and represents that result at its output.

For the most part, inputs to the network are handled in parallel format. Essentially, the parallel nature of the network inputs means that they can be manipulated in any number of ways, such as by integration, differentiation, or selective processing to yield very specific results. The output results can be of two different varieties, one being an output for which no current reference exists, and the other being an expected output.

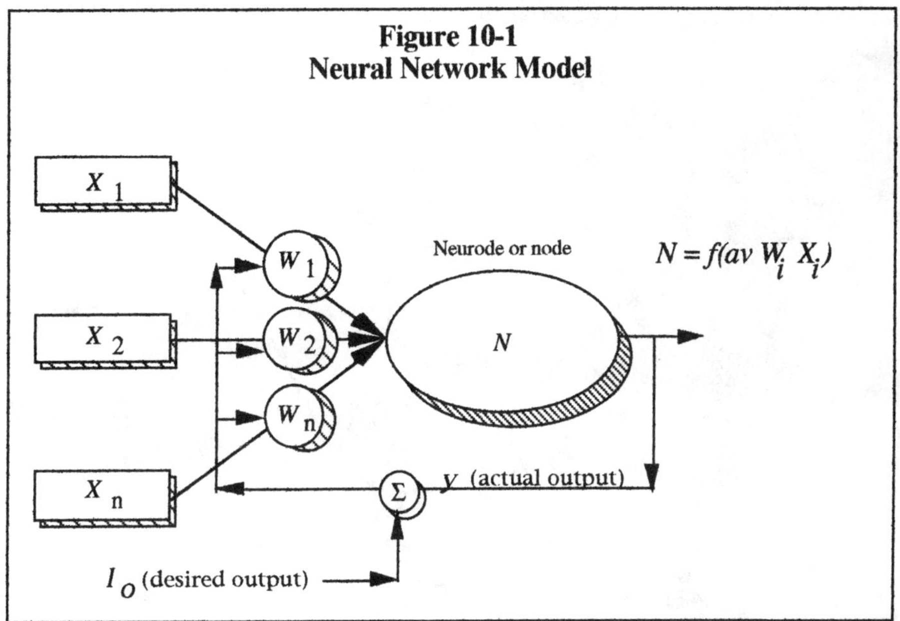

Several repetitions of inputs through the network can be used to modify the weights associated with this network, the end result of which yields a new imprint that becomes a memory of the particular event. This memory is created by the adjustment of the weights after each of several passes through the network. Another output result is one for which a reference already exists. Subsequent input streams can be compared against previously existing weighting factors to determine if measured outputs are consistent with expectations. The results are rarely exact analytical solutions to a problem, but rather fit within or outside some domain of expectation, as will be seen later in Figures 10-7 and 10-9.

Succeeding inputs of similar nature will yield similar output patterns or results. Some input patterns that are close to a particular reference, but not exactly the same, must be examined more closely, and if one is within some threshold for the organism or system, it is declared to be that recognized pattern. If it is not, the system's or organism's neural network must then attempt to associate the received pattern with other stored associations in memory. These associations take the form of substituting a

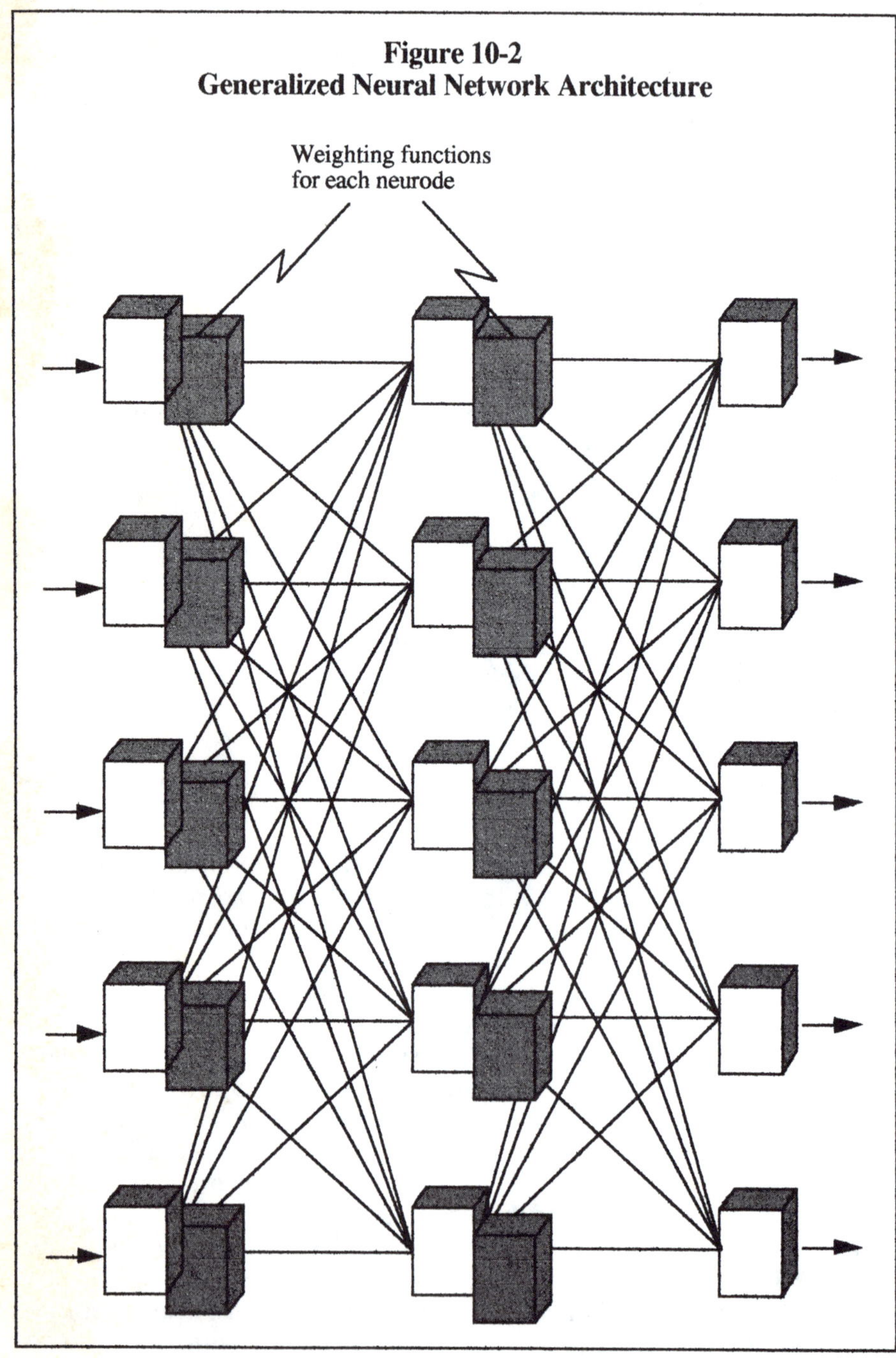

Figure 10-2
Generalized Neural Network Architecture

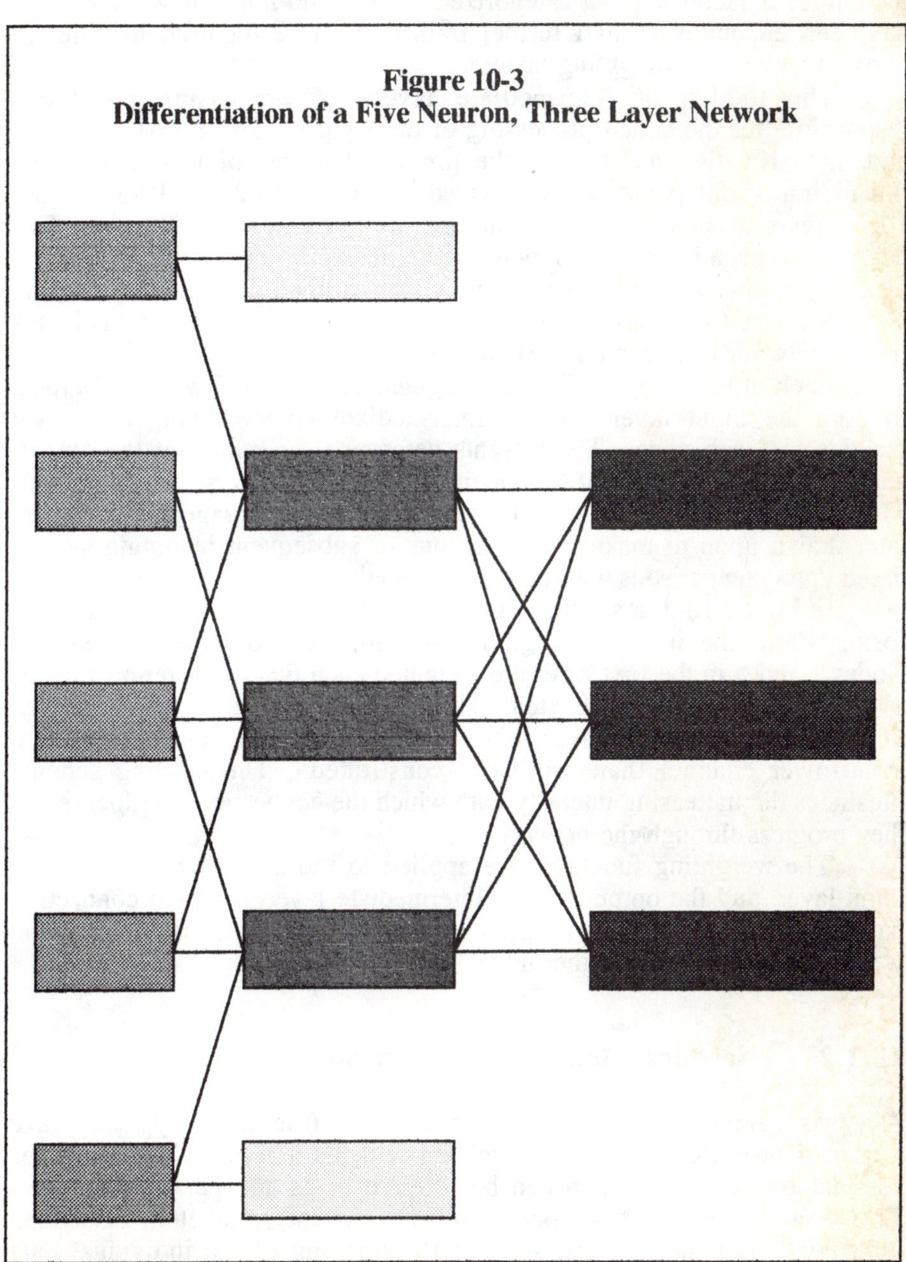

Figure 10-3
Differentiation of a Five Neuron, Three Layer Network

new set of weighting factors within the system's repertoire to determine if some set may fit the inputs encountered. If these weighting factors do fit, the output is identified and categorized. If they do not, a new experience has been encountered, and further training will be required in order to derive a new set of weighting factors.

The hidden or intermediate layers of processing are those responsible for the actual processing of the signals. Signal differentiation and intensification are two of the process features of this layer. An illustration of this paradigm is provided in Figures 10-2 and 10-3. Figure 10-2 shows a network of 15 neurodes arranged in three layers of five neurodes to each layer. The inputs to the network may come from any meaningful source such as a machine vision system, a signal receiver, the proprioceptive (position) sensors of a robot, or the operational status inputs from a network management system. [4]

Each input to Figure 10-2 is weighted as illustrated with the outputs of both the input layer and the intermediate layers having their own weights for each node. The weights determine the nature of the output, and therefore the imprint of the incoming stimulus to the succeeding layer. These imprints may be retained in some type of storage medium, and later drawn upon to make interpretations of subsequent incoming signals based upon comparisons with the stored signals.

In Figure 10-3, a specific example of processing is shown in abstract form, where the incoming signals are subjected to different weights. Nodes 1 and 5 in the first layer are weighted such that the outermost nodes at layer 2 are essentially negated. The second-layer signals are reinforced in the third layer, and the overall effect is to focus the incoming signal into a narrower channel than originally constituted. The shading scheme illustrates the increasing intensity with which the connections are focused as they progress through the network.

The weighting functions are applied to the outputs of the first, or input layer, and the outputs of the intermediate layer(s). Each connection has its own weight assigned. Each connection carries the same value and weighting function to all other nodes from that neurode.

10.1.2   Network Management Scenario

A network element, such as a smart multiplexor (mux), will typically have a normal operational pattern associated with its activity. The attributes associated with this pattern can be referred to as an operational vector. This vector consists of the pertinent and relevant parameters associated with this smart multiplexor, such as the tracking of the individual data streams currently accessing this mux, and the error rates being experienced across its available channels. During normal activities, the operational vector exhibits a normal pattern of activity through the values associated with its attributes. This normal pattern, if applied to a neural network, will provide two types of information.

(1)   If the normal pattern is being initially presented to the neural network, the vector represented will cause weighting functions to be adjusted, so that the network can learn what a normal pattern consists of.

(2)   If the normal pattern is being repeatedly presented to the neural network after training has occurred, no adjustments will be made to the weighting functions.

Thus, the neural network makes a self-adjustment to the incoming vector's normal pattern within its neural network monitoring system. The weights associated with this normal pattern are tuned, and the subsequent vector inputs, presented at intervals, will yield results within acceptable ranges of tolerance. Weight adjustments will normally not be made after initial training. Anomalous conditions will trigger outputs that fail to meet expectations.

An adjustment in the system architecture, such as the number of communications channels, or parameter ranges, such as a increase in acceptable error rates, will cause a new training requirement and subsequent adjustments in the weighting factors involved. Now two sets of weighting factors are in existence, and the system has more flexibility than before because the neural network can accommodate differing sets of acceptable conditions. In other words, no error signals are generated for conditions within these ranges. A detected anomaly, however, results from one or more of the input vector attributes failing to meet standards, assuming there is more than one set of input values. The normal smart mux operational vector might be represented as follows:

$$\text{mux}_{\text{smart input}} = V_n (X_{1n}, X_{2n}, X_{3n},...., X_{nn})$$

where:

$$V_n = \text{the normal operational vector for the smart mux}$$
$$X_{nn} = \text{one of several attributes associated with the}$$

vector under consideration, such that the first n represents the value of the attribute, and the second n represents the fact that the attribute is within normal limits.

The anomalous condition now received may result in the following vector:

$$\text{mux}_{\text{smart input}} \quad = \quad V_a (X_{1n}, , X_{3a}, ..., X_{nn})$$

where:

$$X_{na} = \quad \text{an input attribute that is out-of-limits, or in error}$$

Input parameters, X_{2a}, X_{3a}, are now displaying abnormal behavior and will result in an anomalous output for the neural network. The output pattern for the network no longer matches the expected results, and an alarm may be generated. In addition, other patterns may be resurrected from a storage medium to compare results in an effort to isolate the problem using other patterns that have different problems already embedded in them.

10.1.3   Key Design Issues

There are several key issues associated with the design of neural networks that will be discussed here. These key issues are: (1) the requirements for the number of input neurodes; (2) the weighting of the input signals; (3) the number of intermediate neurodes in the net; (4) the transfer functions associated with the intermediate nodes; (5) the number of output neurodes; (6) the method of feedback training used to adjust the weighting factors at the intermediate and output layers of the net; and (7) the physical connectivity between the nodes within a layer and those between layers. Each of these will be expanded upon below.

(1)   Input Nodes

The input nodes receive the input signals presented to each of them and pass these signals without alteration and weighting to the intermediate layer. These nodes serve to organize the inputs for latter processing. The question is how many input nodes are needed to accommodate the incoming data stream. The answer depends upon the number of data points being processed at any particular time. For a signal being received over the ether, each input node might represent a sample of that signal, either time-based or frequency-based. This could yield a fairly manageable number of input nodes. For a more complex situation such as an image, each pixel might be assigned to an input node, or a set of pixels, averaged together, might be assigned to an input node. Each input signal is also normalized to a value between 0 and 1. The equations describing the inputs to and outputs from the input layer are shown below.

Net input$_{\text{input layer}}$ $= i_i$

Net output$_{\text{input layer}}$ $= i_i$

where:

i_i $=$ input signals to first, or input, layer

(2) Weighting the Input Signals

The intermediate layer receives the input signals' components from the input layer. Each signal being presented to the intermediate layer is weighted such that some function between -1 and +1 is multiplied with the input signal to form a compound value. Thus, each input has its own assigned weight. Each of these is, in turn, summed and applied to each of the intermediate layer nodes, or neurodes. This condition can be expressed as follows:

Net input$_{\text{intermediate layer}}$ $= \sum w_1 \cdot o_i$

Output$_{\text{intermediate layer}}$ $= o_j = 1 \Big/ [1 + \exp{(-i_j)}]$

where:

i_j $=$ input to the intermediate layer

$\exp$ $=$ e

o_j $=$ output of intermediate layer

w_1 $=$ weighting function for intermediate layer

o_i $=$ output of first, or input, layer

(3) Number of Intermediate Nodes Required

The number of intermediate nodes required for processing and training can be variable. The number depends upon several factors, such as the application and the number of inputs, but as a general rule, it has been found that the number required is somewhat of a fudge factor. This number is approximately the square root of the number of inputs, plus the number of outputs, plus a few extra nodes. How many *a few extra* ends up

being is not defined with any precision. This definition is left to the
designer and user.

$$\text{No. } n_{il} \quad = \quad \left(\Sigma i_j \right)^{1/2}$$

where:

$$\Sigma\, i_j \quad = \quad \text{summation of the number of inputs}$$
$$\text{No. } n_{il} \quad = \quad \text{number of nodes required in the}$$
$$\text{intermediate layers}$$

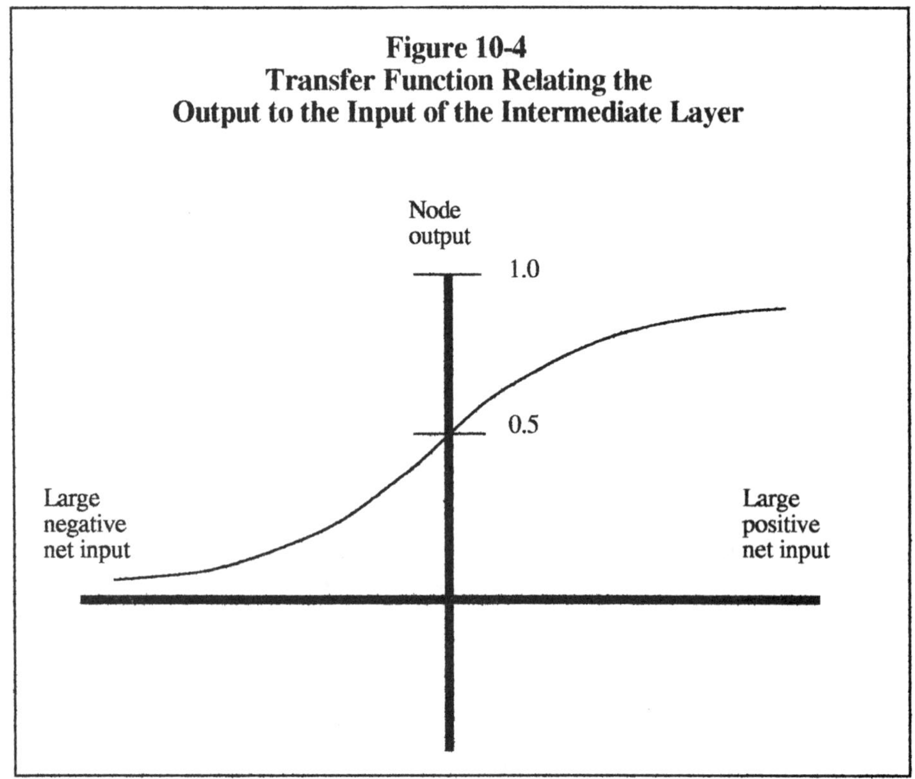

Figure 10-4
Transfer Function Relating the
Output to the Input of the Intermediate Layer

(4)   Intermediate Layer Transfer Function

The output of each intermediate layer node is determined by what is
referred to as a transfer function. This function maps the relationship of
any output based upon the value of the input. It could be linear; i.e., for
any input, the output bears a direct and consistent relationship to the input,

as, for example, a linear relationship based upon the equation $y = mx + b$, where m is a constant slope for any value of y given x. Besides being linear, the transfer function could also be exponential, geometric, or any of a number of other choices. A considerable amount of work has been performed using the sigmoid, or S-shaped, function. It is represented as shown in Figure 10-4. Any transfer function could be used that has a derivative at any point along its function.

(5)   Number of Output Nodes

The number of output nodes required is more straightforward than the determination of input nodes. It is the number of data classifications needed by the system to define the memory reference. Such output numbers might be 525, if the number of outputs are processing each pixel per line of a TV picture, or 72, if the number of outputs are the points of a 72-point Fast Fourier transform (FFT). The general rule is that each node can represent each point or set of points that are *mutually exclusive*. Points that are dependent upon each other may be corrupted if forced to be combined with other points that are dependent in the same ways. Thus, biases may add together and artificially affect each other.

(6)   Physical Connectivity

Current research into the nature and efficacy of neural network configurations has led to certain approaches for physical interconnection. [6] The input layers are many times arranged in linear or rectangular patterns. Linear patterns help with the filtering processes, and rectangular configurations help to globalize the incoming signals.

(7)   Method of Training

Training is accomplished by feedback into the system. The problem arises as to how such feedback is defined and conveyed to that system. Training, which is a function of feedback, is realized by error adjustments. There are two principal approaches to the training problem, one being to compare and adjust, as necessary, the network weights after each training pattern is determined from a pass through the neural network. This is referred to as on-line, or single-pattern, training. In this scenario, a single pass of signal inputs is made through the network. The results yield a pattern that can then be fed back through the network to adjust the weights at the intermediate and output node layers. This process is repeated after each training pattern passes through the network.

The other training approach is to collect the errors over the training period, in an average, summed, squared-error accumulation. The formula for the total pattern error is:

$$\text{Error}_{\text{pattern}} = 0.5 \sum (\text{value}_{\text{pattern}} - \text{value}_{\text{output}}) \Big/ N_{\text{patterns}}$$

where:

$\text{value}_{\text{pattern}}$	=	current value of the pattern
$\text{value}_{\text{output}}$	=	current value of the output
N_{patterns}	=	number of patterns

The single node, or neurode, error value is expressed as follows:

$$\sigma = (\text{value}_{\text{pattern}} - \text{value}_{\text{output}}) \, \text{value}_{\text{output}} (1 - \text{value}_{\text{output}})$$

where:

$\text{value}_{\text{pattern}}$	=	node pattern value
$\text{value}_{\text{output}}$	=	node output value

(8). Weight Adjustments for Training

Each node has at least one output, and may have several inputs. For each of these, there is a weight assigned as depicted in Figure 10-1. The output of each node may be at variance with expected or required outputs. This yields an error signal that is fed back to the weighting functions, and adjustments are made using an equation that reweighs the original weighting functions. This relationship, known as the delta function, is described as follows:

$$W_{\text{new}} = W_{\text{old}} + \beta E X \Big/ |X|^2$$

where:

W_{new}	=	new weight
W_{old}	=	old weight
E	=	output error = $\text{output}_{\text{desired}} - \text{output}_{\text{actual}}$
β	=	learning constant
X	=	input vector

$$|X| \qquad = \qquad \text{magnitude of input vector}$$

(9) Firing Thresholds for Nodes

The node, as, for example, the nodal paradigm illustrated in Figure 10-1, may have a multiple set of inputs. The minimum value of the inputs necessary for the node, in turn, to trigger an output response, is called its threshold. If, for instance, the weighted inputs cause an incremental increase in total value of 0.8 each, and the threshold is set at 2.0, then at least three inputs must be registered at the node for it to relay an output. Additionally, some outputs may be inhibitory, i.e., they may cause nodes to produce negative output values. These negative outputs are then fed as inputs to other nodes and affect their total trigger values accordingly.

When each of the weights is reevaluated based upon the relationships shown above, an adjustment is made to the old value, and that weight is recalculated for application to the next pass. Figure 10-1 shows that the actual output of each node, or neurode, is measured against the desired output of each. The error resulting is calculated as shown above and the weights are adjusted accordingly. This error is used to modify each weight associated with that node.

10.1.4   Neural Processing Examples

Certain examples of neural processing can make the neural network more understandable. [5] Suppose that a sonar problem is posed that consists of three possible outcomes, namely rocks, fish, and mines. Clearly, mines are the most compelling of the three items, because of their possible danger. The feature space associated with this problem might be illustrated as shown in Figure 10-5. In real life, the feature space would probably consist of more than just two possible inputs; however, the two-dimensional nature of the example provides a better understanding for illustration purposes. The feature space has two possible inputs, namely F1 and F2. Each of these are supplied to an intermediate node where the F1 and F2 inputs are summed. This node, in turn, provides its output to a third node where the F1, F2, and summation of F1 and F2 are again summed. This output now describes a set of spaces that, in some combination, identifies the features present.

This result is shown in Figure 10-6. The combination of the inputs, F1 and F2, when weighted in varying ways, results in areas within the feature space that are candidates for the objects of interest. These feature space areas are illustrated in Figure 10-7. These areas outlined in different shades are derived from iterations of signals passing through the network, and being fed back to adjust the weights associated with each node.

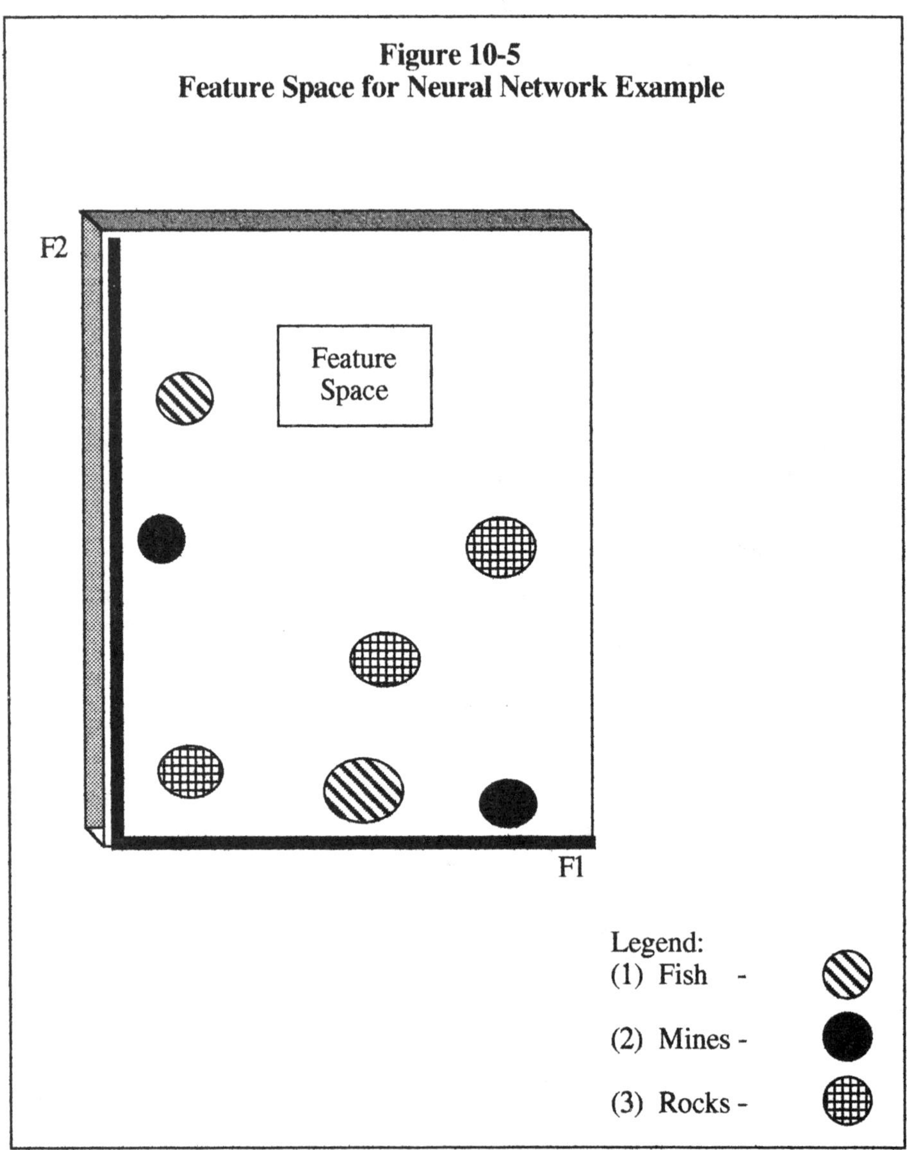

In a network management situation, all, a majority, or the most important monitoring points may be presented to the neural network for analysis and resolution. [2] Such a simplified example is provided in

Figure 10-8. Here a very modest network is replicated where two inputs
are monitored, namely the signal-to-noise ratio between a microwave

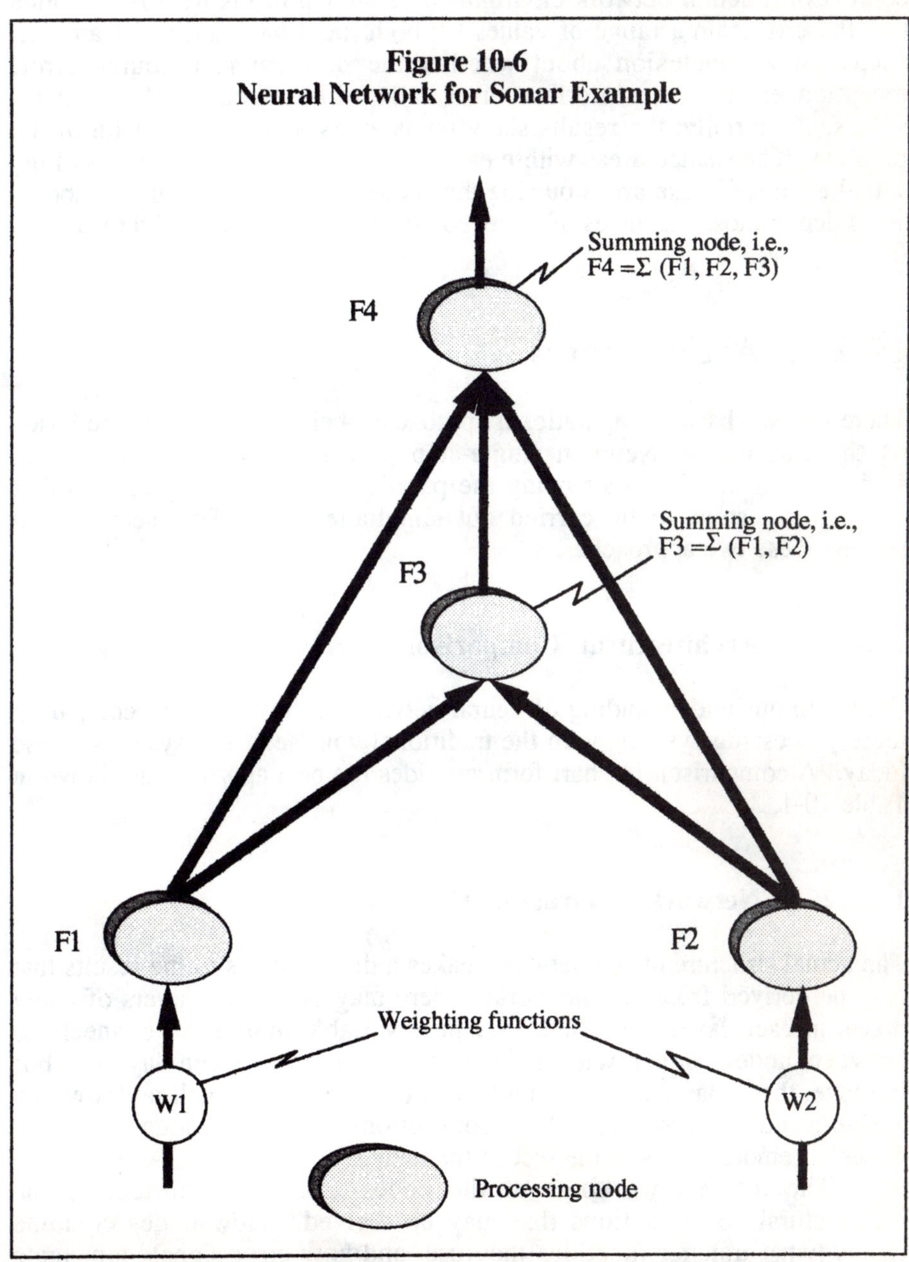

antenna link, and the block error rate for a packet assembler/dissembler (PAD) interface. A depiction of how this example can be monitored in the context of a neural network environment is shown in Figure 10-9. Notice that there is again a range of values for both input parameters, F1 and F2, such that a conclusion about whether the problem is a source error, reception error, or an interface error falls within a range of parameter values. Pictorially, the results show up as areas within the domain of the problem. The shaded areas within each labeled area are the instances of the actual events. Those areas outside the shaded areas but within the labeled areas depict those instances of correct analyses that yield no valid results.

10.2   Architectures

There are two basic computational approaches being used today, one based on the classic von Neuman single-step processing and system control architecture, and the other being the parallel processor architectures that allow many steps to be carried out simultaneously. This section will address these two approaches.

10.2.1   Architectural Comparison

We begin our understanding of neural network architectures by comparing these processing systems with the traditional von Neumann systems in use today. A comparison in chart form provides the best approach as shown in Table 10-1.

10.2.2   Network Characteristics

The actual structure of the network makes a difference as to the results that may be derived from that network. There may be several layers of many nodes in each layer, and there may be a variable number of connections between nodes, either within the same layer or between layers. For instance, there may be a full complement of connections to all nodes within a layer, and a sparse number of connections between a node and other nodes at remote points in the rest of the network.

The internal workings of the nodes are also important to the architectural configurations that may be derived. How nodes combine inputs, what transfer functions they use, and how error signals propagate through the system are all important issues that determine the operational characteristics of the network and, therefore, what its architecture should be to support such characteristics.

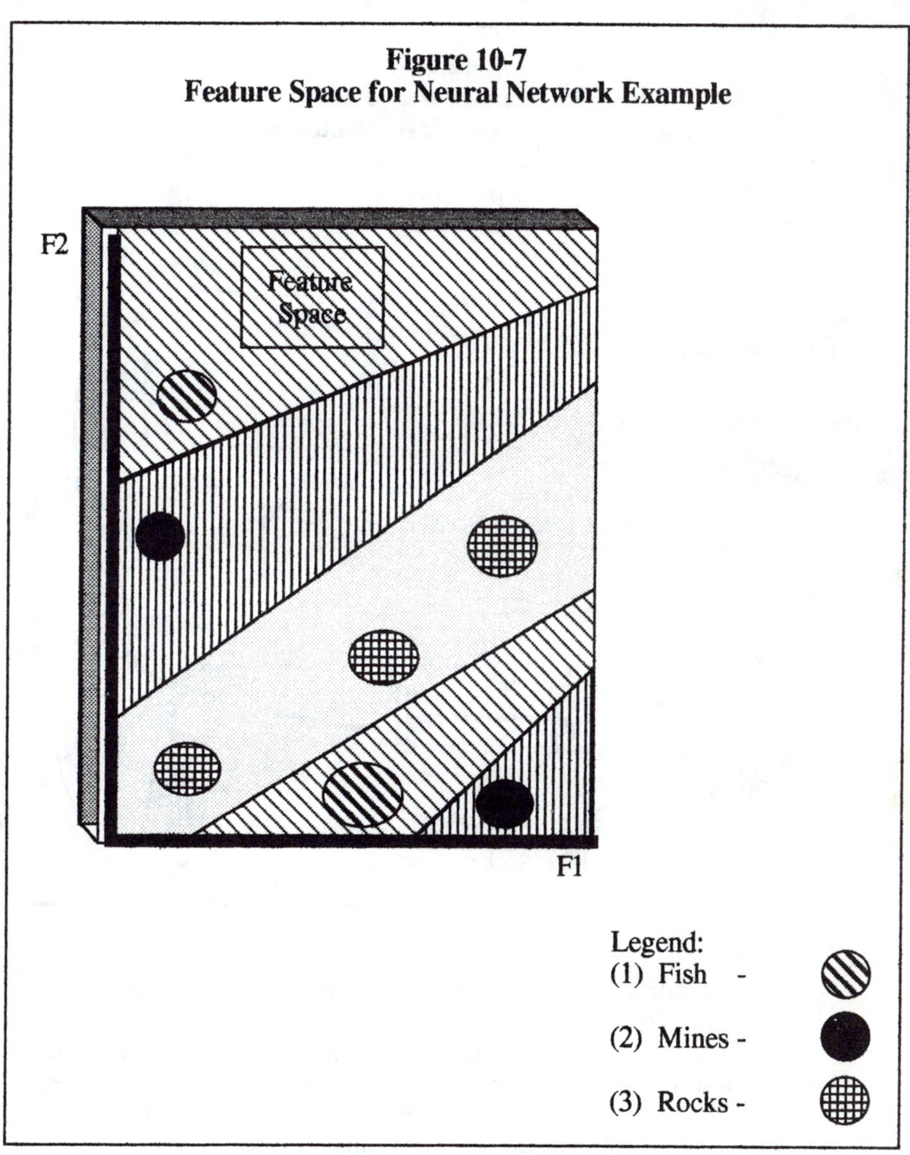

Figure 10-7
Feature Space for Neural Network Example

The training methods, whether they be externally or internally supervised, are also important. The former type may be easier to implement, but recurring costs may be an aggravation. The latter type may be more difficult to implement initially, but possibly will be much less expensive in the long run because of its ability to self-correct.

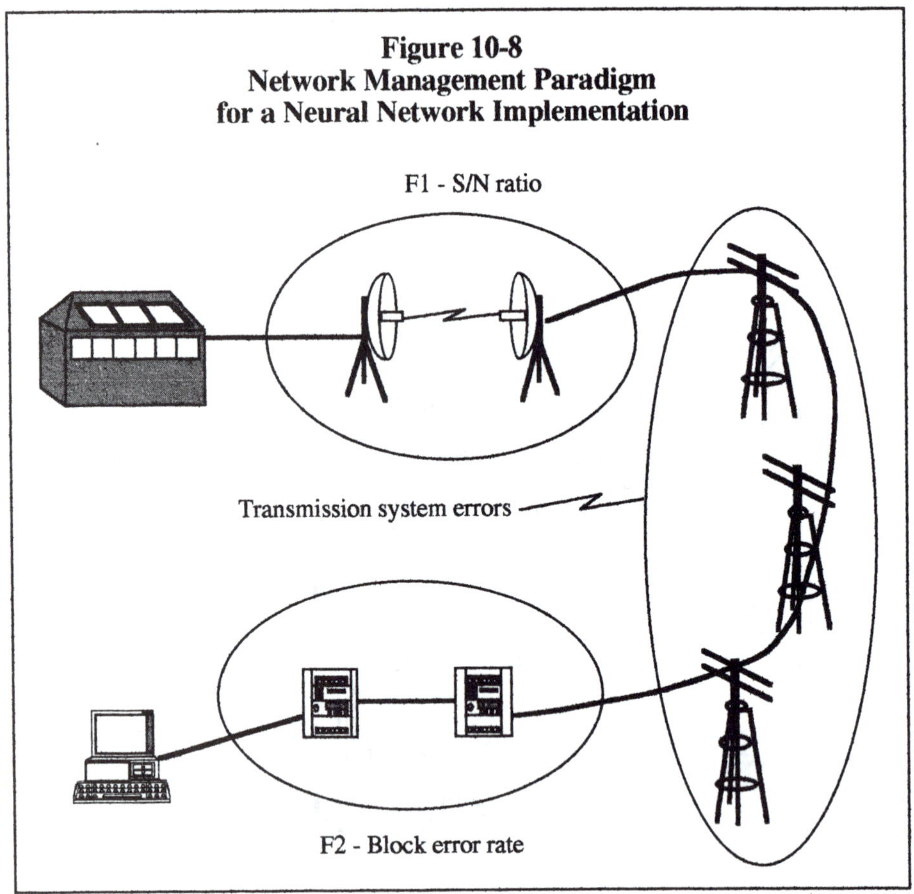

Self-organization has also been a recurring theme in work to date. This objective has focused upon a set of self-organization networks known as Kohonen networks. These are characterized by just two layer configurations, foregoing the intermediate layer. Each input node, or neurode, has the same criterion as the basic architecture described above; namely, each input node is assigned to a segmented part of an input signal. Within the input layer, each of these nodes is, in turn, connected to each and every node in the output layer. Additionally, each of these connections is associated with a weighting function.

Each of the layers can be represented in a variety of ways, but typically, each layer is represented by a two-dimensional array of nodes with the layout resembling a rectangle or a hexagon. These layered configurations are referred to as *slabs*. In some cases there may be two or more slabs in a layer. An example of the concept of the self-organizing network is shown in Figure 10-10.

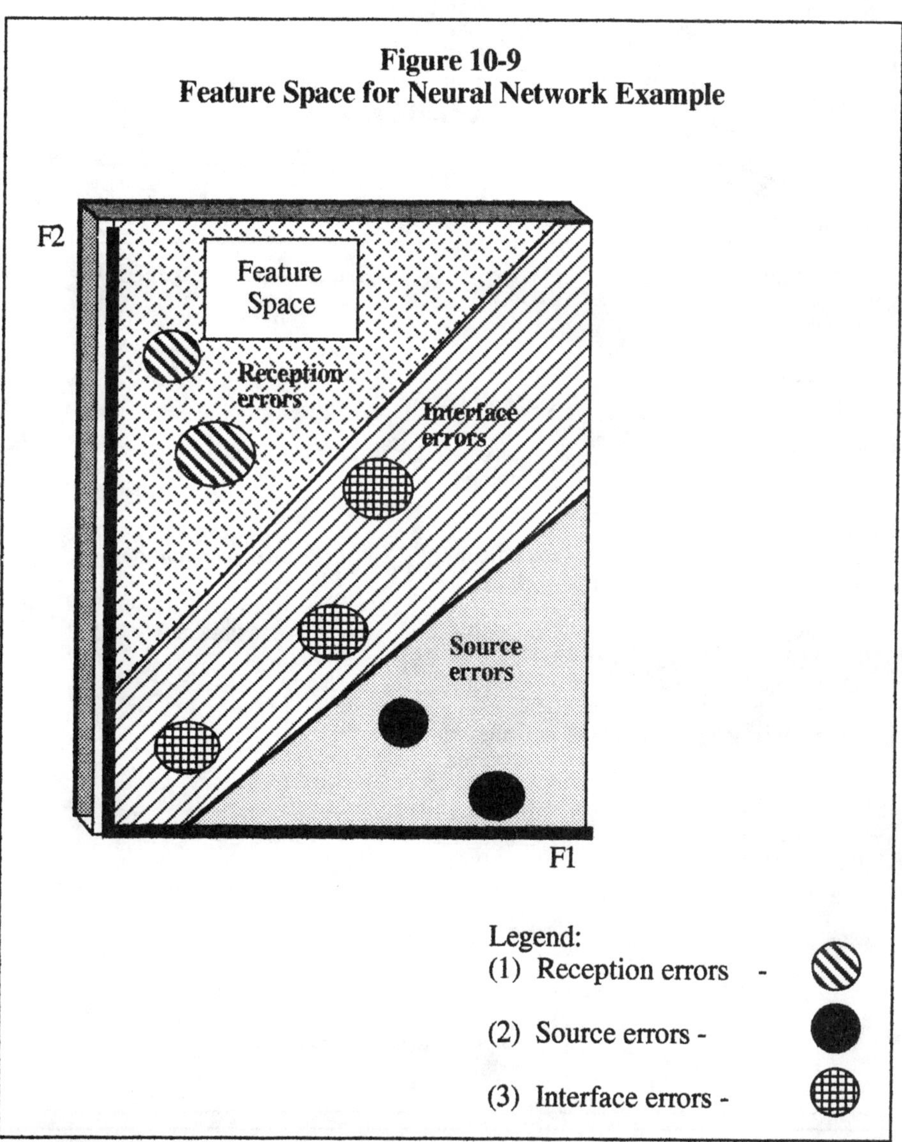

    In this figure, the layers consist of only one slab with nine nodes in each slab. As usual, each input signal segment is relayed through an input node. Each of the nodes is, in turn, connected to each of the output nodes, with an appropriate weighting function applied to each connection to the output layer. This is a rectangular arrangement. In situations where the

Table 10-1
Comparison of Neural Versus Conventional Computational Systems

Comparison Category	von Neumann	Neural Networks
Processing units	one	Large numbers, e.g., 1000 or more
Complexity	High	Low
Level of processor intreconnect	None or few	High
Memory design	Discrete storage	Diffuse storage
Hardware technology		
Current	Silicon ICs	Software on conventional CPUs
Future	Optical ICs	ASICs and optical ICs
Development state	Mature	Infancy
Problem solving approach	Rule-directed	Trained
Source of knowledge	Input data	Trial and error
Advantages	Precise	Does not rely upon rules
	Well-developed technology	Ability to deal with fuzzy logic
Disadvantages	Explicit rules required	Inexact results

layer may consist of more than one slab, each slab is assigned to process similar signal components, or attributes from the same source. These separable inputs are passed to separate slabs in the output slab as well. Each connection between the input and output nodes carries its own weighting function, as illustrated in the basic model in Figure 10-11.

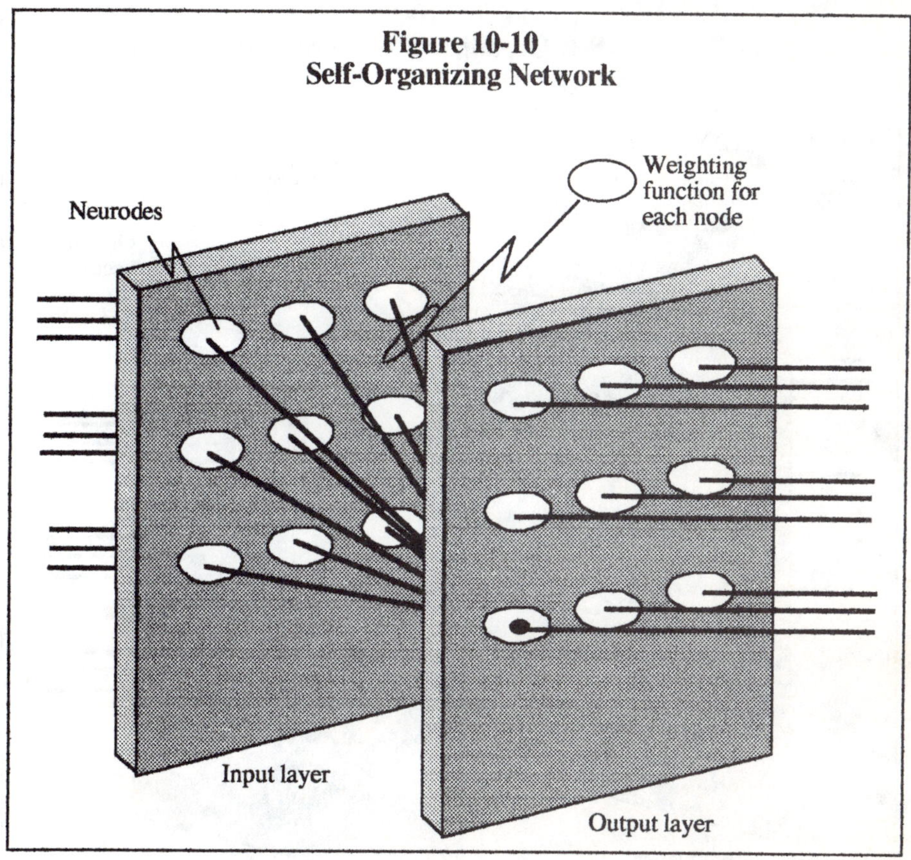

10.3 Implementations

10.3.1 Constraints

Commercial development and introduction of neural networks for today's problems presents several types of constraints, some of which are major

and some of which are fairly small challenges. In essence, the technology
is currently available for some rudimentary applications. Practically all of
the current applications rely upon software simulations, with the current
state of progress lagging the promise of the capabilities inherent in neural
networks until hardware developments catch up.

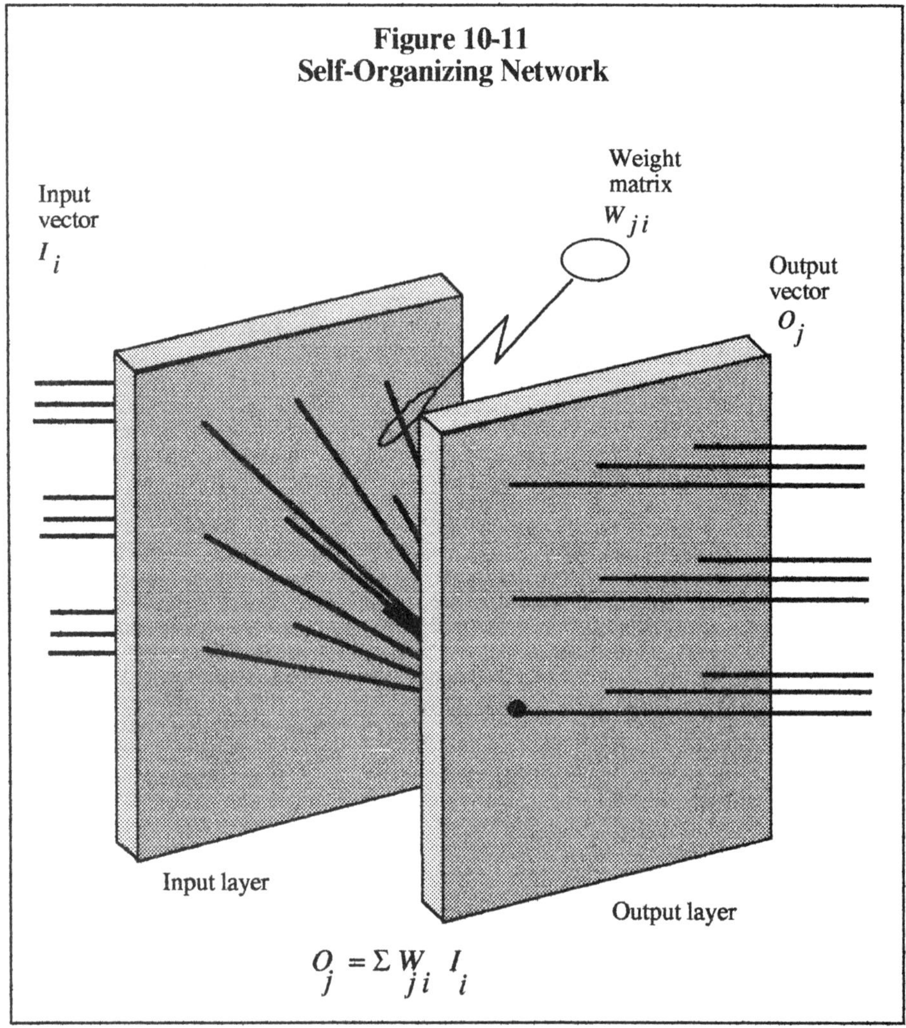

Most of the current applications require considerable efforts at
preprocessing prior to introduction into the neural network. The
implication of these early observations is that neural nets may be part of

more complex solutions to technology problems, as opposed to stand-alone solutions. [3]

Two major requirements for the future are that neural networks must be capable of converging on solutions within certain numbers of training rounds. Secondly, they must be able to explain their actions as part of larger complex handling situations.

The implementation of a suitable environment for neural network solutions is a three-phase activity. Currently, software is the approach of choice, mostly because there are few if any other solutions at present. Next will come ASICs (application-specific integrated circuits) that will embody many node-like processing points. Finally, will come the neurocomputer that will be VHSIC (very high speed integrated circuits), and/or opto-electric, or biological technology, chips capable of accommodating a very large number of nodes.

10.3.2   Demand

Neural networks are likely to receive widespread demand in the commercial sector in coming years. Such application industries include manufacturing, processing, financial, and engineering areas. Specific applications involve machine vision, signal processing, natural language, and robotics. Neural networks seem to have significant advantages over conventional computing represented by numeric manipulations, transactions, and rule-based actions.

Summary:

The architecture and design of neural networks is based upon the use of simple building blocks referred to as neurodes. These elements are rudimentary in their construction, but when combined with many other similar neurodes are capable of complex operations, heretofore thought only to be associated with human intellect, such as trial-and-error goal attainment behavior. The goals established for the neural network systems are attained through layering the processing attack upon the problem as well as weighting functions associated with each of the neurodes. Training these networks is an essential part of their successful usage against complex problems. This is the mechanism by which the weighting functions are derived. Certain classes of problems, for example undersea object identification, have validated the use of even simple neural networks to analyze, by comparison of responses, objects requiring identification.

References

(1)   Caudill, M., *Neural Network Primer*, Miller Freeman Publications, San Francisco, 1990.

(2)   Goyal, S. K., and R. W. Worrest, "Expert System Applications to Network Management," in *Expert System Applications to Telecommunications*, Liebowitz, J.(ed.), John Wiley, New York, 1988, p 1.

(3)   Harris, C. J., and I. White (eds.), *Advances in Command, Control and Communication Systems*, Peter Peregrinus Ltd., London, UK, 1987.

(4)   Henning, W., "Bus Systems," *Sensors and Actuators A*, 25-27 (1991), pp 109-113.

(5)   Hoffman, M., "Technology Profile Neural Networks," *Techmonitoring*, SRI International, July, 1991.

(6)   Lemmon, A., *Marvel - A Knowledge-Based Planning System*, GTE Laboratories, Internal Report, 1986.

Index

h
habituation 279
hard failure 90
hash technique 15
heap sort 16
heuristics 142, 218
hidden layer 294
hierarchial clustering method 127
hierarchical searching 225
high-definition television (HDTV)
    286
homogeneity 97
human reasoning 172
human-machine interface 148, 157
hypothesis testing 116

i
if-then-else statements 212
implications of the problem 154
in-band signaling 115
indepth fault mode analysis 215
indifference 55
indirect data fusion 112
inferencing
    inference methods 122
    inference cycle 180
    inference engine 152
information model 216
inhibitory neuron 271
input layer 294
integrated access 5
integrated systems 20
integrity 62
intelligent
    intelligent database systems
    241
    intelligent shells 250
    intelligent systems (IS) 243
interaction scenarios 247
interactive menu selections 246
intermittent faults 60
interneurons 273
ISDN 286

isolation of the problem 153

k
key word matching approach 192
knowledge
    knowledge acquisition 222,
      227
    knowledge base design 207
    knowledge blackboards 184
    knowledge czar 182
    knowledge degradation 175
    knowledge engineer 188,
      211
    knowledge engineering 204
    knowledge frames 183
    knowledge networks 183
    knowledge representation
      227
    knowledge scripts 183
    knowledge sharing 174
    knowledge specificity 224
    knowledge-based systems
      approach 192

l
layered problem solving 225
learning 210, 282
limited logic systems approach
    192
line replaceable units (LRUs) 230
linear compensation 96
linearization 95
linguistic variables 124
linkage approach 129
LISP 14
local area network (LAN) 25
LU 6.2 160

m
machine translation 192
machine vision 208
Macintosh 161
maintainability 231
maintainability design 210

About The Author

Larry L. Ball, Ph.D. has had a wide range of experiences in both the commercial and government sectors. He has practiced as an engineer, businessman, scientist, and entrepreneur. His working experience includes telecommunications systems engineering, satellite systems design, medical imaging product management, and software systems development. Dr. Ball holds degrees from the United States Air Force Academy (BS), California State University at San Jose (MS), The University of Southern California (MS), and Stanford University (Ph.D.). He has served as manager of JPL's Deep Space Network, started his own medical systems company, built a software development organization that performed on expert systems as well as realtime control programs, designed telecommunications systems, and has managed medical imaging product development. He is currently chief engineer for MPR, Teltech, Ltd., Burnaby, B.C., Canada, a division of B.C. Telephone. His previous book, _Cost-Effective Network Management_, gives insights into how the network management functions cxan be performed more efficiently. Dr. Ball brings insight from many business areas and technologies to the development of this technical reference, _Smart Systems for Network Management_.